DISCOVERY OF THE LIFE-ORGANIZING PRINCIPLE

DISCOVERY OF THE LIFE-ORGANIZING PRINCIPLE

In Search of the Fundamental Laws of Life

E. M. ELSHEIK

iUniverse LLC
Bloomington

DISCOVERY OF THE LIFE-ORGANIZING PRINCIPLE
In Search of the Fundamental Laws of Life

iUniverse books may be ordered through booksellers or by contacting:

iUniverse LLC
1663 Liberty Drive
Bloomington, IN 47403
www.iuniverse.com
1-800-Authors (1-800-288-4677)

Because of the dynamic nature of the Internet, any web addresses or
links contained in this book may have changed since publication and may
no longer be valid. The views expressed in this work are solely those
of the author and do not necessarily reflect the views of the publisher,
and the publisher hereby disclaims any responsibility for them.

Any people depicted in stock imagery provided by Thinkstock are
models, and such images are being used for illustrative purposes only.
Certain stock imagery © Thinkstock.

ISBN: 978-1-4917-2717-1 (sc)
ISBN: 978-1-4917-2719-5 (hc)
ISBN: 978-1-4917-2718-8 (e)

Library of Congress Control Number: 2014904323

Printed in the United States of America.

iUniverse rev. date: 03/15/2014

DISCOVERY OF THE LIFE-ORGANIZING PRINCIPLE

In Search of the Fundamental Laws of Life

E. M. ELSHEIK

iUniverse LLC
Bloomington

iUniverse books may be ordered through booksellers or by contacting:

iUniverse LLC
1663 Liberty Drive
Bloomington, IN 47403
www.iuniverse.com
1-800-Authors (1-800-288-4677)

Because of the dynamic nature of the Internet, any web addresses or links contained in this book may have changed since publication and may no longer be valid. The views expressed in this work are solely those of the author and do not necessarily reflect the views of the publisher, and the publisher hereby disclaims any responsibility for them.

Any people depicted in stock imagery provided by Thinkstock are models, and such images are being used for illustrative purposes only. Certain stock imagery © Thinkstock.

ISBN: 978-1-4917-2717-1 (sc)
ISBN: 978-1-4917-2719-5 (hc)
ISBN: 978-1-4917-2718-8 (e)

Library of Congress Control Number: 2014904323

Printed in the United States of America.

iUniverse rev. date: 03/15/2014

*To my mother, to all women, to all those who struggle
for a better human future full of love and peace*

CONTENTS

PREFACE

We accept the mainstream biology and mainstream physics claim that life is a physical rather than supernatural phenomenon. Then why do living systems self-organize, self-replicate, and self-evolve while nonliving systems do not? According to the second law of thermodynamics, a living system should decay and disintegrate; however, this does not occur only when the living system dies. Why? The answer is that the living system is far from an equilibrium thermodynamic open system that exchanges matter-energy with the surroundings. In doing so, it sustains its living state without decomposition. The question, then, is how can it be possible for the system to traverse a path of increasing complexity from thermodynamic equilibrium to maintain a state far from equilibrium thermodynamics? What is the driving force? Is it an intelligent designer?

According to quantum mechanics, a living system, being a macroscopic localized system, should be decoherent and lack useful energy for its function. On the contrary, the living system is coherent and rich in useful energy, so what is the source of its coherence? The dynamics of a physical system is embedded in phase space coordinates from which the system's equation of motion can be derived. On the contrary, the living system dynamics depends on its bioinformation or biocomplexity rather than on the space coordinates it occupies. So how can it be possible to discover the life-organizing principle that contains the dynamical essence of a living system irrespective of the phase space coordinates? Moreover, if biological evolution is not a random process but subject to the life-organizing principle, then it has to have a goal function or target criterion. What is it?

It is clear that the problem of the nature of life is neither purely biological nor purely physical in the ordinary sense; it is both. To resolve such a problem, four steps are necessary:

First: We propose a paradigm shift that broadens the concepts of information, life fractal nature, quantum field, and least-action principle.

Second: Based on the paradigm shift, we discover what physically distinguishes life from nonlife (i.e., the genome's capacity to generate bioinformation oscillations through successive generations).

Third: Based on the bioinformation oscillations, we formulate the life-organizing principle, which is a generalized Schrödinger type of system with vitality, a measure of bioinformation, as path variable.

Fourth: Based on the life-organizing principle, we derive the first and second laws of self-organization, which explain biological evolution and development. Moreover, they generate functional genetic code capable of instructing viable proteins and determine conclusive biological evolution goal function, which is maximization of total vitality.

Finally, the genome's total bioinformation generates two survival components: reproductive fitness component and total vitality fitness component. Thus evolution as maximization of total vitality implies maximization of creativity and altruism (faeeliya). This extension of Darwinian

theory substantiates the theory of multilevel selection and facilitates a more dynamic conception of human nature. It facilitates a transition to postcapitalism society, as capitalism is underpinned by a transitional phase of evolving human nature. It also reconciles the existing contradiction between science and the final goals of religion.

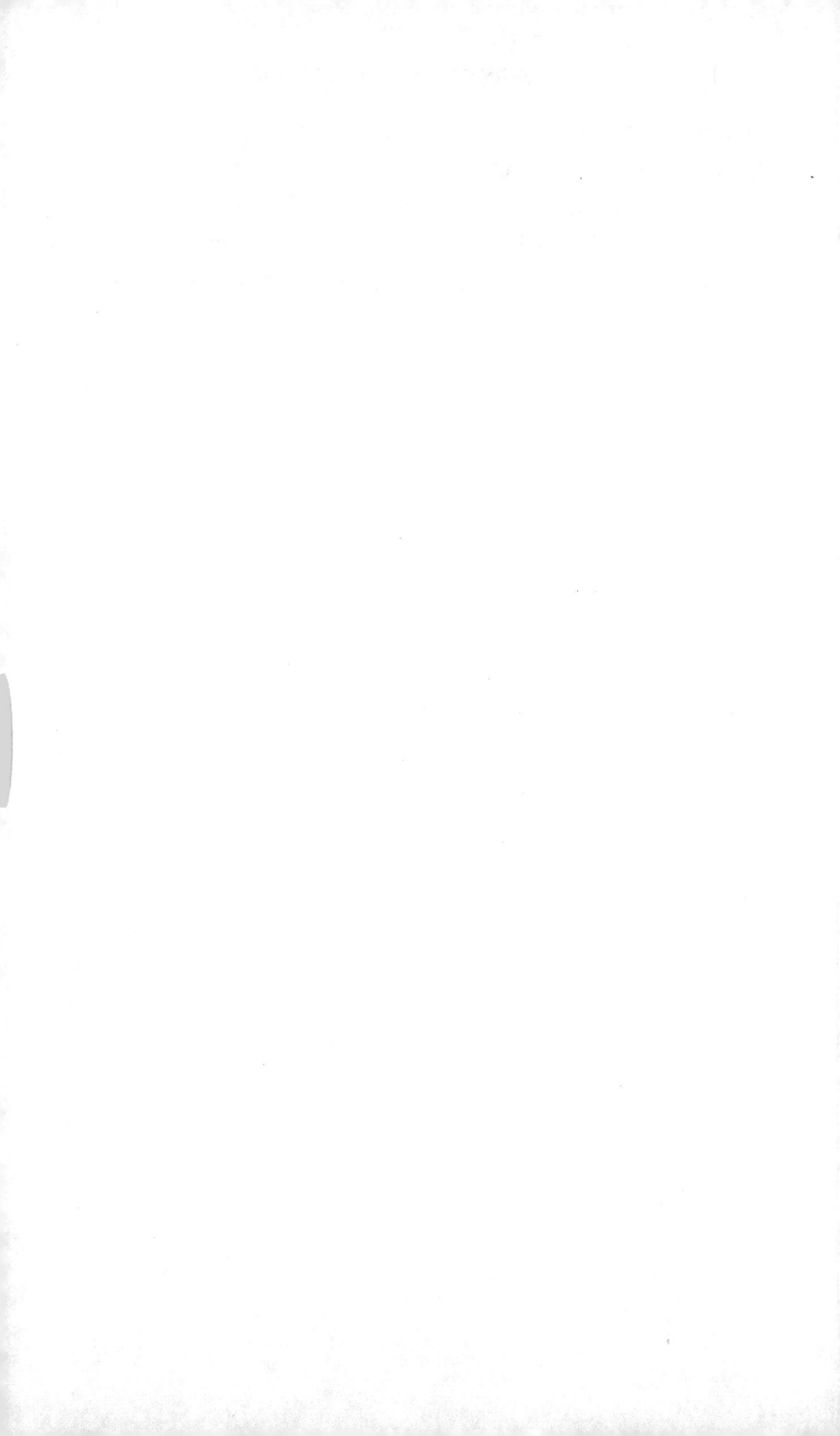

ACKNOWLEDGMENTS

I would like to acknowledge my indebtedness to Dan Brooks, Robert Ulanowicz, Sven Jorgensen, and the late M. O. Taha for their helpful comments and support. Thanks also to anonymous referees whose critical comments helped me to improve my theory. I'm also grateful to Dan Winter for collaboration and support.

I would like to thank Dr. Siddig Umbada, Dr. Nour Eldeen Abd Elrahman, Mr. Hassan Omer Ibrahim, Dr. Abd Alla Abdeen and Elnour A. Ali for their financial support to publish the book. Thanks are also due to iUniverse for kind treatment and for the encouraging discount that helped me to publish this work.

INTRODUCTION

What is life? There's no doubt that this is the most challenging question the human mind has ever encountered. Living systems are problematic in the sense that although they are composed of the same elements found among inanimate systems and do not violate the laws governing the physicohemical transformations and reactions of these elements, they grow, develop, and evolve, thus generating bioinformation or biocomplexity in a peculiar manner unattainable to inanimate systems. So how can we explain the peculiar behavior of living systems? Do we regard them as complicated machines, subject to the same physicochemical laws governing their elementary inanimate constituents and transformations? Do we regard them as autonomous systems that obey new laws that are independent of physics? Is there a life-organizing principle of a physical nature that does not belong to ordinary physics (i.e., inanimate physics)? In short, what is the nature of life?

I have been working on the last alternative with sincere dedication for more than forty years. Such an alternative necessitates discovering a fundamental physical property that distinguishes life from nonlife. If such a property does exist, that means it has escaped human endeavor and scientific imagination throughout human history. I dare say that I have discovered this property, which broadens the ontological foundation of contemporary physical theory and reveals the secret of life. Thus the secret of life resides in the DNA or genome as a quantum information fractal field that generates bioinformation oscillations through successive generations.

The bioinformation is a measure of biocomplexity that is developmental functional complexity and increases before adulthood, having a maximum when the organism is fully grown. It decreases afterward and becomes zero when the organism dies. Since such behavior is periodic or oscillatory for successive generations, and since the bioinformation has the dimensions of energy and information, a new quantum information fractal field theory can be developed in order to account for biotic evolution and development.

How I envisaged this idea and how I kept developing it despite extremely challenging circumstances and enormous sacrifices is itself a lesson on the psycho-existential roots of creativity and perseverance, for the realization of life as "activeness and effectiveness" is nothing other than an abstraction of my personal life experience. It is the experience of a little kid who lost his father and was lucky to have the opportunity to develop himself through his own initiative, morally, intellectually, and academically. The personal experience of every individual is part of the universe, so it possesses a universal element. Consequently, creativity is the recognition, extraction, and unfolding (generalization) of the universal element embedded in individual experience. Having been able to mold and develop my personality and psychology in accordance with what are regarded as virtuous human values and ambitions through acquiring knowledge, later, when I came to realize that human knowledge is in crisis (i.e., we don't know what human nature is and don't know what life is), my compass lost direction. These problems became not only epistemological problems but also deep psychological and existential problems. In other words, there is no way to be in harmony with myself and with the world unless I find or develop a coherent conceptualization of what being human is and what life is. The identity crisis of human knowledge concerning the nature of life and human nature became my personal identity crisis.

This is why I started studying biology and philosophy while I was actually a mathematics student in the Department of Mathematics, Faculty of Science, University of Khartoum, where I was studying physics and mathematics in the late sixties of the last century. Later on, I did my master's degree thesis on "Biological Evolution—A. Nonlinear Irreversible Quantum Approach," and my informal supervisor was internationally respected Sudanese physicist M. O. Taha. Then having a grant from the International Center for Theoretical Physics, Trieste, Italy, I joined Carlos Leguizamon's research group at the University of Buenos Aires, 1987. From there, I had the opportunity to contribute to the Third International Congress of Bio-mathematics, Santiago, Schile, 1987, with a paper titled "Evolution of Unicellular Organisms in Terms of Vitality Function." The paper has been cited by some scientists. In fact, since the seventies of the last century, no editor in chief of the *Journal of Theoretical Biology or BioSystems* hasn't encountered versions of my proposed new theory at least once. The persistent rejection of my work followed by critical comments from anonymous referees was a main source of inspiration for developing and improving the presentation of my work.

In fact, it was always evident to me that what these honorable scientists criticized and regarded as inadequate was not the order of nature I envisage but the way I presented such order. I also know that they themselves have no answer to the problem under consideration. Not only that, but I think it is difficult (if not impossible) to resolve the problem of life based on one specialty (a biologist being specialized in a certain branch of biology or a physicist specialized in a certain branch of physics). Life is a whole, and it has to be approached holistically—that is, on a broader ontological basis than what a specialist is usually equipped for. This is why I was sure I was on the right track, and the critique was a source of enrichment to me rather than a source of disappointment.

Moreover, I received helpful comments from famous scientists, including Brian Goodwin, H. H. Pattee, Bob Ulanowicz, and Sven Jorgensen. Finally, my paper, "Toward a New Physical Theory of Biotic Evolution and Development," was published in the peer-reviewed international journal *Ecological Modeling*, 2010. A main postulate of the theory is that living systems are information-generating systems. They are capable by themselves, on naturalistic basis, to generate information, and they do not need an intelligent designer. After that, I made important developments to the theory—interestingly, not by adding external elements but mainly by extracting information hidden in the theory. Thus a conclusive credential of the theory is that it is mechanistic, deductive, and predictive, and almost all its predictions are experimentally falsifiable.

I have also developed what I call faeeliya analysis. Faeeliya is an Arabic word that means activeness and effectiveness; however, as a term, I use it to mean creativity and altruism. Hence, faeeliya analysis is a method for revealing the faeeliya of individuals, societies, and literary texts. Faeeliya represents the social aspect or human dimension of the proposed new theory. I have four books (in Arabic) concerning faeeliya analysis. Four students had their master's degrees in literature and philosophy using faeeliya analysis; a fifth student had a PhD in literature from Elfatih University, Libya, 2009, using faeeliya analysis. In fact, Faeeliya as creativity and altruism is not restricted to humans. It is a fundamental attribute (as I try to show later) of biological evolution.

Puzzled over the nature of life, scientists either claim that the question of life and its mechanisms is insoluble, as something the human mind could never penetrate and understand, or they give phenomenological accounts by enumerating the different characteristics or hallmarks that designate a system to be alive, such as metabolism, heredity, evolution, information system, and so forth. However, there are certain obstacles that generate chaos in understanding the nature of life and block the discovery of the life-organizing principle. They are as follows.

Thermodynamic Barrier:

Pross (2003) asserts that life's far-from-equilibrium state is central to the dilemma of how biology and physics interrelate. From a purely thermodynamic perspective, living systems do not violate the second law of thermodynamics. Analogous to a refrigerator, which can transfer heat from cold to hot in the reverse direction of the natural one, and can do so through the consumption of energy, a living system can maintain its far-from-equilibrium state through the continual utilization of energy. "But how could such a highly organized energy-gathering entity have come about?" Pross (2003) asked. In order for a living system to attain far-from-equilibrium state, it has to traverse a path of maximum action or equally well a path of minimum entropy, so what is the driving force and what is the mechanism to overcome such a barrier?

Quantum Decoherence Barrier:

Quantum theory describes the evolution of delocalized objects (i.e., objects subject to the wave function). Such objects, under certain conditions, may exhibit coherence, which is the collective cooperation of a large number of particles in a single quantum state. In this regard, living systems are macroscopic localized decoherent objects that operate beyond the realm of quantum mechanics. Nonetheless, Davies (2004) is concerned with whether there are nontrivial quantum phenomena relevant for biology. Nontrivial means the presence of long-ranged, long-lived, or multiparticle quantum coherences, the explicit use of quantum entanglement; and so on. He emphasized, "If quantum mechanics is to play a nontrivial role in bio-systems, then some way to sustain quantum coherence at least for biochemically, if not biologically, significant time scales must be found. Without this crucial step, quantum biology is dead" (Davies 2004). Currently, photosynthesis, the process of vision, the sense of smell, or the magnetic orientation of migrant birds, is considered nontrivial quantum phenomena relevant for biology. These phenomena, even if regarded as nontrivial, raise the question of whether they justify reduction of biology to ordinary physics or whether they reflect the ability of biological organization to harness physical phenomena. If living systems are capable of harnessing various physical and chemical phenomena, what is the source of their apparent coherence and functionality? How does life transcend the decoherence barrier?

Phase Space Barrier:

In physics, the phase space coordinates of the system at any given time are composed of all the systems dynamical variables. Because of this, it is possible to calculate the state of the system at any given time in the future or the past, through integration of Hamilton's or Lagrange's equation of motion. Longo (2012) emphasized that in biological evolution, the phase space itself changes persistently, in ways that cannot be prestated. So he argues that no law in the physical sense entails the evolution of life. Life evolution being independent of phase space coordinates or the underlying physical fields raises the question of whether or not it is subject to a new type of physical field with different dynamical variables. This also raises the question of what physically distinguishes life from nonlife.

Reductionism Barrier:

Reductionism claims that a living system is nothing but the sum of its parts, and that an account for it can be reduced to accounts of its individual constituents. This can be said of objects, phenomena, explanation, theories, and meanings. Reductionism is underpinned by the doctrine of causal closure or completeness of physics. The stronger formulations of causal closure assert this: "No physical event has a cause outside the physical domain" (Jaegwon Kim 1998), or "Physical effects have only physical causes" (Agustin Vincente 2006). That is, the stronger formulations assert that for physical events, causes *other* than physical causes do not exist. In this perspective, living systems having physical effects entail that biology is reducible to ordinary physics.

On the other hand antireductionism claims that a living system is irreducible to physics because the living system is a whole that is greater than the arithmetical sum of its parts. Thus Pattee (1968) asserts that if living and inanimate systems are subject to the same set of physicochemical laws, then the question of why living systems grow and evolve while the nonliving systems do not is not answered.

In response to this dilemma, Schrödinger (1944) says, "We must therefore not be discouraged by the difficulty of interpreting life by the ordinary laws of physics. For that is just what is to be expected from the knowledge we have gained of the structure of living matter. We must be prepared to find a new type of physical law prevailing in it. Or are we to term it a non-physical, not to say super-physical, law?" He then confirms that the new law or principle is nothing more than the principle of quantum theory all over again.

To overcome these obstacles in order to discover the nature of life and its organizing principle, or what Schrödinger calls "the new principle which is not alien to physics," we propose the following paradigm shift.

Broadening the Concept of Information

It is important to distinguish between physical information as a measure of physical complexity that is static and bioinformation as a measure of biocomplexity that is developmental and functional complexity. So it is appropriate to describe bioinformation or biocomplexity in terms of information dimension as well as energy dimension. For this sake, Elsheikh (2010) defined vitality, which is the genome capacity to generate developmental functional complexity as a function of the organism's genome physical information, total matter-energy metabolized by the organism, and its life expectancy. Vitality increases before adulthood, has a maximum when the organism is fully grown, decreases afterward, and becomes zero when the organism dies. Considering a unicellular organism that divides for successive generations, vitality becomes a periodic function of time. Thus it represents the bioinformation oscillations generated by the genome through successive generations.

Identification of Life Golden Ratio
Based Fractal Nature:

A fractal is a rough or fragmented geometric shape that can be subdivided into parts, each of which is approximately a reduced size copy of the whole. As Mandelbrot (1982) indicates, a fractal is generally self-similar and independent of scale. Fractals model complex physical processes and dynamical systems. The underlying principle is that a simple process that goes through infinitely much iteration becomes a very complex process. Kurakin (2011), Knott (2001), Wille

(2012), Perez (1990), and Winter (2012) have all emphasized the ubiquitous nature of fractality in biological hierarchy, and that Fibonacci numbers or the golden ratio is the basis of the natural fractal design of living systems. In particular, Winter (2012) demonstrates that the DNA, having dodecahedron geometric structure, is a golden ratio optimized fractal field that has the electric property of generating recursive constructive interference (i.e., fractal field phase conjugation).

Broadening Quantum Field Theory:

The discovery of the above-mentioned bioinformation oscillations generated by the genome for successive generations encourages us to envisage a new type of quantum field. Since the quantum field is defined as a function over space and time, the new field is defined as a function over bioinformation and time. The new field is nothing other than the outcome of the collective dynamics of ordinary physical fields optimized by golden ratio based DNA fractal geometry. Thus the genome is a quantum information fractal field (QIFF) that generates, in addition to weak electromagnetic waves, bioinformation oscillations through successive generations. Consequently, matter has complementary properties: matter waves, microscopically, at high mass density, and bioinformation oscillations at high information density. Based on this complementarity, a material system, animate or inanimate, does not simultaneously possess both matter waves and bioinformation oscillations descriptions. Thus both reductionism and antireductionism are partially correct: life is not reducible to ordinary physics, yet it is reducible to a new generalized physics, quantum information fractal field theory, QIFFT.

Extension of Least-Action Principle:

The least action is an extremum principle according to which a mechanical system behaves in such a way that the action S (time integral over the Lagrange function) is minimized. Action has the dimensions of energy and time. Now it is clear that as an organism grows and develops, its action increases rather than decreases. To account for such an observation, there must be a maximum action principle, according to which a system spontaneously traverses a path of maximum action. According to the least-action principle, a conservative system traverses a path of least action because its rate of change of action decreases, while according to the maximum action principle, a nonconservative system traverses a path of maximum action because its rate of change of action increases. Based on the maximum action principle, the organism's rate of change of action is proportional to its developmental functional complexity (that is, bioinformation). Hence the maximum action principle is the driving force of self-organization and bioinformation generation.

Based on this paradigm shift, we propose the following postulates of a new quantum information fractal field theory (QIFFT):

[i]   A living system's genome is a self-organizing, self-replicating and self-evolving quantum information fractal field, QIFF.

[ii] The QIFF generates, in addition to weak electromagnetic waves, bioinformation oscillations through successive generations.

[iii] The bioinformation oscillations contain the dynamical essence of the living system.

[iv] The bioinformation sustains the living state.

We claim that these postulates are appropriate to define life and provide bases for its mathematical representation. In particular, we claim the discovery of the following:

The Life-Organizing Principle: It is a generalized Schrödinger type of system with vitality that is a measure of developmental functional complexity (bioinformation) as a path variable. The life-organizing principle (LOP) is a nonlinear, nonconservative, and irreversible second order differential equation that facilitates the system structural stability (i.e., dynamic stability or an attractor). It is a minor attractor when describing cell dynamics and a major attractor when describing multicellular organism dynamics. A cell type is an example of a minor attractor that belongs to the basin of a major attractor. Hence both phylogeny and ontogeny are processes that generate and assemble minor attractors. The phase of the life-organizing principle represents a maximum action principle from which the first law of self-organization can be derived. Moreover, assuming the living system has no life before birth and has no life after death as boundary conditions, the second law of self-organization can be derived on the basis of the life-organizing principle.

The First Law of Self-Organization: This states that the organism's rate of change of action is proportional to its developmental functional complexity (bioinformation); it increases before adulthood, has a maximum when the organism is fully grown, decreases afterward, and becomes zero when the organism dies. Based on the maximum action principle, life's self-organization does not need an intelligent designer; the nature of intelligent design is nature's intelligence.

The Second Law of Self-Organization: This states that total vitality, which is biological evolution fractal goal function or target criterion, is directly proportional to Fibonacci numbers, genome's physical information and varies inversely with respect to the organism's frequency of bioinformation oscillations. Total vitality is the product of the organism's total action, life span, and genome physical information. The Fibonacci numbers characterize the organism quantum functional stationary states (states of dynamic stability); they represent the states of beneficial mutational changes phylogenetically or the states of organ or pattern formation ontogenetically. In other words, they represent minor attractors states ontogenetically and major attractors states phylogenetically.

Conservation of Genome's Total Bioinformation: This law states that the organism total natality density function plus its total vitality is conserved. The conserved quantity facilitates the derivation of organism and population growth functions; it equally well accounts for the apparent lower natality rates characteristic of more evolved living systems. Moreover, the genome's total bioinformation generates two survival components—reproductive fitness component and total vitality fitness component—which paves the way for a theory of multilevel selection.

Ecosystem Dynamics: Here we propose mathematical formulation for Jorgensen's ecological law of thermodynamics. The law states: "Ecosystem development in all phases will move away from thermodynamic equilibrium and has the propensity to select the components and the organization that yields the highest flux of useful energy throughout the system and the most exergy stored in the system. This also corresponds to the highest ascendancy" (Jorgensen 2006).

A Unified Theory of Life: Based on the proposed quantum information fractal field theory, life (genome) is a quantum information fractal field that generates autonomous self-sustained bioinformation oscillations for successive generations. However, a specific individual living system is said to be alive if its rate of change of action is greater than zero and its vitality is greater than zero. In other words, life (genome) is a quantum information fractal field subject to the first and second laws of self-organization. Such a theory is capable to account for all life forms, ontogenetically as well as phylogenetically.

Moreover, the QIFFT proposes a new foundation of human knowledge; on the one hand, it extends the concept of matter in order to account for life phenomenon and reconciles the contradiction or disparity between physics and biology by discovering the QIFF. On the other hand, it extends Darwinian theory by extending the concept of survival and revealing reproductive fitness and total vitality (or faeeliya) fitness components of survival, which support the theory of multilevel selection. Hence, the new extended Darwinian theory reconciles the disparity or contradiction between biology and the humanities, and it provides theoretical basis for understanding the origin of morality and religion. It also provides theoretical basis for a transition from capitalism to postcapitalism society, as capitalism is underpinned by a transitional phase of evolving human nature.

CHAPTER 1

WHAT IS LIFE?

1.1—CHARACTERISTICS OF LIFE

Physicochemical Bases

During the nineteenth century, both chemists and biologists had been able to discover the elementary constituent of their subject matter. Biologists showed that the cell is the elementary constituent from which an organism is built and that all cells arise from preexisting cells. Chemists first focused their attention on inorganic matter and found that it consists of a small number of different atoms in definite proportions. However, when they came to analyze organic substances, the picture seemed quite different. Two substances might have exactly the same composition yet show distinctly different properties. For example, ethyl alcohol is composed of two carbon atoms, one oxygen atom, and six hydrogen atoms; so is dimethyl ether, yet one is liquid at room temperature while the other is a gas. Moreover, organic molecules contain many more atoms than the simple inorganic ones, and there seemed to be no reason the way they were combined.

Thus J. J. Berzelius (1779-1847) and other chemists decided that the chemistry of life was something apart that obeyed its own set of subtle rules and that only living tissue could make an organic compound. However, in 1828, German chemist Friedrich Wohler, a student of Berzelius's, produced an organic substance in the laboratory by heating ammonium cyanate compound, which was then generally considered inorganic. This spectacular discovery paved the way for all the miracles that biochemistry and (later on) molecular biology have been witnessing since then. Biochemists now have at their disposal all sorts of synthetics: explosived, insecticides, plastics, fibers, rubbers, and proteins.

Living matter contains elements, inorganic and organic compounds. The most abundant elements found in protoplasm are oxygen, carbon, hydrogen, and nitrogen. Other elements found in small quantities are calcium, phosphorus, chlorine, sulfur, potassium, sodium, magnesium, and iron. Some other elements not generally present that may occur in small quantities are lithium, boron, fluorine, aluminum, silicon, vanadium, manganese, cobalt, copper, zinc, selenium, bromine, molybdenum, cadmium, iodine, and barium. The nonmetallic elements oxygen, carbon, hydrogen and nitrogen are the most abundant in living matter because of their contributions to the formation of organic molecules. The metallic elements commonly constitute only a small fraction of the cell volume but are essential to many phases of cell metabolism.

There are no inorganic compounds in protoplasm that do not occur in the same form in nonliving matter. The most abundant of these is water, which constitutes 75 percent of living matter. Carbon dioxide is present as a dissolved gas. Also dissolved in the water of cells are salts, acids, and bases. Carbon dioxide is utilized by green plants in food synthesis and released in all cells as an end product of food breakdown. Many inorganic compounds in living matter are common salts, acids, or bases. When dissolved in water, their molecules tend to dissociate into ions, positively charged cations from hydrogen and the metallic elements and negatively charged anions from the nonmetallic elements.

Table salt, for example, ionizes into sodium ions that are positive (Na^+) and chlorine ions that are negative (Cl^-).Acids invariably form hydrogen ions (H^+) and anions that differ with the kind of acid. Bases yield various types of cations and invariably hydroxyl ions (OH^-).

A great many complex organic compounds occur in living matter: carbohydrates, lipids, proteins, and nucleic acids. These organic molecules are composed of a relatively small number of kinds of molecular subunits. The large molecules break down into their component units on hydrolysis (digestion), and synthesis occur through the reverse process of bonding by dehydration (condensation). Sugars and the products of their condensations (polysaccharides) are carbohydrates. A sugar consists of a carbon chain generally arranged in a ring structure, flanked by hydrogen atoms and hydroxyl (OH^-) groups, the hydrogen and oxygen being present typically in the same proportions as in water. Many carbohydrates are hexose (6—carbon) sugars or yield hexose when digested. Pentose (5—carbon) sugars occur in nucleic acids and certain polysaccharides and are important in photosynthesis. The initial carbohydrate manufacture takes place in organisms in the process of photosynthesis. The energy stored, which comes from the sun, is then available for biological work through the oxidation of the carbohydrates that are produced. The carbohydrates are classified as monosaccharides, disaccharides, and polysaccharides. Monosaccharides are simple sugars like glucose ($C_6 H_{12} O_6$), one of the key molecules in cellular respiration, and ribose ($C_5 H_{10} O_5$) a subunit of nucleotides, which form nucleic acids.

Disaccharides, double sugars, consist of two linked monosaccharides. Examples: sucrose ($C_{12} H_{22} O_{11}$), table sugar, and maltose ($C_{12} H_{22} O_{11}$), derived in the digestion of starch. Sucrose is formed of one molecule each of glucose and fructose; maltose is formed of two molecules of glucose. Polysaccharides are formed from numerous linked monosaccharides. Example: starch, cellulose, and glycogen all have the same empirical formula ($C_6 H_{10} O_5)_x$.

The lipids are chemically diverse molecules that are generally insoluble or only slightly soluble in water but are soluble in ethyl alcohol, ether, and some other organic solvent. Three types of lipids commonly occur in protoplasm:

- Fats (true fats or neutral fats) are compounds that consist of one molecule of glycerol and three molecules of fatty acids. Each fatty acid consists of a long hydrocarbon chain that is saturated if the carbons contain the maximum possible number of attached hydrogen (as in palmitic acid) and unsaturated if any carbon atoms are double bonded, resulting in less hydrogen per carnon (as in oleic acid). Fats occur primarily as food reserve molecules, formed when food material is abundant and degraded when food is in demand.

- Phospholipids are similar to the true fats, but one of the fatty acids is replaced by a phosphate group, usually with additional water-soluble molecules.

- Steroids are structurally different from true fats, having four fused carbon rings with an attached carbon chain of varying length. In small concentrations, various forms of these compounds exert biological regulatory effects on sexual development and function as well as on certain aspects of metabolism in higher organisms. Common examples include cholesterol, vitamin D, sex hormones.

Proteins:

These are large molecules consisting of long unbranched chains of amino acids. Although the twenty or so different types of amino acids have distinctive structural characteristics and therefore distinctive chemical properties, they all have the same configuration at one location. This configuration, an amino group $(-NH_2)$ and an acidic carboxyl group (-COOH) attached to the same carbon atom, is the key to the manner in which adjacent amino acids are linked by peptide bonds. These bonds are formed between the amino group of one amino acid and the carboxyl group of the next, one molecule of water being removed in the formation of each bond. The remaining components of each amino acid extend from the protein as side chains (arms) consisting of hydrogen (as in glycine, the simplest amino acid) or of a variety of straight chain or ring compounds, which may also contain sulfur or additional oxygen and nitrogen.

Almost all cellular differences within an organism and among organisms can be traced to protein differences. Proteins are essential in the formation and maintenance of the structural and functional machinery of cells. They are after deamination (removal of $-NH_2$), a source of energy in the diet being oxidized much as carbohydrates are. There are different types of proteins:

i. Proteins that function in structural capacity (e.g., keratins of fingernails, skin, and hair as well as collagens of connective tissue)

ii. Proteins that are hormones, regulators of metabolic processes (e.g., insulin)

iii. Proteins that are components of chromosomes (e.g., histones)

iv. Proteins that function in the transport of oxygen (e.g., hemoglobin)

v. Proteins that are organic catalysts (enzymes) controlling reaction occurrence and rate (e.g., enzymes of digestion and enzymes of cellular metabolism). Like other catalysts, they activate a reaction, but they are not used up in the process.

Nucleic Acids:

Nucleic acids are organic compounds that are even larger than proteins. As with proteins, they are composed of repeating subunits. In nucleic acids, these subunits are nucleotides. Each nucleotide consists of a nitrogen-containing organic base (chemically a purine or purimidine) a pentose (5—carbon) sugar (ribose or deoxyribose), and phosphoric acid. There are two types of nucleic acids:

DNA (deoxyribonucleic acid), which has deoxyribose in the nucleotides, and RNA (ribonucleic acid), which has ribose as a nucleotide component. Each type of nucleic acid has four possible nucleotides, depending on which nitrogenous base is present. DNA nucleotides contain the purines guanine and adenine and the pyrimidines cytocine and thymine. The same kinds of nucleotides compose RNA, except that thymine is replaced by uracil. The sugar of each nucleotide is attached to the adjacent nucleotide through the phosphate, and the purine and pyrimidine are side branches.

The DNA molecule consists of two parallel nucleotide chains in a double helix, with the organic bases extending toward each other. These bases pair adenine only with thymine, and guanine only with cytosine, with stabilizing bonds (hydrogen bonds) holding the adjacent polynucleotide chains in position. As a result, the strands are complementary but not identical. DNA is predominantly found in chromosomes, whereby it performs basic functions in the control of cell expression and in the transmission of heredity information from one cell to the next cell generation. The theory that the pair of complementary polynucleotides that composes the DNA molecule separates, forming new DNA molecules after the synthesis of new complementary units, gives a chemical explanation of gene replication. In contrast to DNA, most RNA is found as single strand on a helix. There are three types of RNA found within the cell: ribosomal RNA, transfer RNA, and messenger RNA. In association with DNA, RNA controls protein synthesis.

Metabolism:

A basic characteristic of living systems is metabolism. The phenomena of metabolism include the taking in of various substances by the organism from the external environment, their assimilation and transformation, and elimination of breakdown products of decomposition. All these processes of transformation are attended by a multiplicity of various chemical, mechanical, thermal, and electrical phenomena and by a continuous transformation of energy. The potential energy of complex organic compounds is liberated through their breakdown and is transformed into heat, mechanical power, and electricity.

Synthetic energy demanding reaction of metabolism is called anabolism, and the reverse, breakdown, energy-releasing reaction is called catabolism. Metabolic reactions occurring within the cell require presence of enzymes. Enzymes are biological catalysts, compounds that accelerate chemical reactions, enabling the cell chemical machinery to operate under its relatively low temperature, low pressure, and relatively restricted PH range. An enzyme catalyzes a reaction by reducing the amount of energy necessary to activate it.

Metabolic reactions operate in sequence, the product of one reaction being the substrate of the next and each serving as an important link in the total chain of reactions. The metabolic sequences can be summarized under four categories: photosynthesis, respiration, biosynthesis, and digestion. Photosynthesis and biosynthesis are both anabolic reactions (more energy goes into the reaction than is released), whereas digestion and respiration, on the other hand, are catabolic reactions (releasing more energy than is required for the reaction).

Food production takes place in chlorophyll-containing cells by the process of photosynthesis. The energy involved is from light, and the raw materials used are water and carbon dioxide from the environment. The net effect, chemically, is reduction, leading to the formation of simple carbohydrates and accompanied by the release of free energy. Respiration is a complex series of intracellular reactions, releasing the energy stored in food. The chemical process is oxidation, and the net effect is the reverse of photosynthesis. The complete process requires oxygen; the original raw materials of photosynthesis, carbon dioxide, and water are given off as end products.

In biosynthesis, simple food molecules produced by photosynthesis or obtained in nutrition are combined (condensed) to form larger molecules. Large carbohydrate molecules (polysaccharides), representing stored food condensed from simple sugar. Stored fats of many different kinds are condensed from glycerol (derived from triose sugar) and fatty acids (long chain compounds derived from many combined acetyl groups). The production of the complex macromolecules of nucleic acids and proteins is controlled by the hereditary material. The cell usually manufactures a variety of large molecules through biosynthesis and assembles them into the structures of intracellular units. Thus the complex architecture of the cell is produced and maintained.

Because a cell membrane is impermeable to larger molecules, these larger molecules must be broken down (digested) for transport and or absorption through cell membranes. During the life of the cell, its various molecular components are continuously being built up and broken down. This gain and loss of molecules is dependent on the rates of synthesis and digestion.

Physicochemical bases of life are not only manifested in that living matter is composed of the same elements as nonliving matter but also certain principles basic to an understanding of physics and chemistry are important in biology. Thus, in both living and inanimate systems, energy may change from one form to another—from potential energy to the kinetic energy of heat or movement as well as from light energy to chemical (potential) energy.

Transformation of chemical energy into various forms of biological activity is accompanied by a change of part of the energy into heat, which is dissipated; such transfers are not 100 percent efficient. It follows that while no energy is lost in accordance with the first law of thermodynamics, there is a decrease in useable energy in the system in accordance with the second law of thermodynamics.

Bioinformation Generating Systems:

In the previous section, it was shown that biological systems, in the final analysis, do compose of the same element (atoms) that are found in inanimate systems and that the biophysical and biochemical transformations and reactions comply with physicochemical laws (i.e., they show no violations for these laws). In this section, we consider a process that is characteristically biological and has no counterpart in the inanimate world, namely the process of self-organization, which increases bio-complexity as manifested in the growth and evolution of biological systems. Ontogenetically as well as phylogenetically biological systems increase in biocomplexity, body size, or life span.

The cell is the basic unit of biocomplexity. Basic cellular processes such as cellular growth, cellular differentiation, and evolution are processes of increased biological organization and complexity. However, at this preliminary stage, let us consider DNA replication and protein synthesis, which substantiate the claim that DNA is a self-organizing and self-replicating system.

The complexity of proteins is unmistakable. For instance, the molecule of a blood protein called cerum albumin was found to contain the following amino acids (Asimov 1972): 15 glycines, 45 valines, 58 leucines, 9 isoleucines, 31 prolines, 33 phenylalanines, 18 tyrosines, 1 tryptophan, 22 serines, 27 thereonines, 32 cystines, 4 cesteines, 6 methionines, 25 arginines, 16 histidines, 58 lysines, 46 asartic acids, and 80 glutamic acids, a total of 526 amino acids of 18 different types built into a protein with molecular weight of about 69,000. All of these 526 amino acids are arranged in a definite order to constitute a definite structure. Clearly a cell cannot make such an arrangement randomly; there must be a deterministic process for generating this biocomplexity.

Eden (1966) presented some simple calculations to demonstrate the degree of improbability to generate useful proteins randomly. If we consider an average protein (word) to consist of 250 amino acid residues (letters), each letter chosen from alphabet of 20 amino acids. There are about 20^{250} such words or 10^{325}, and the probability of choosing a single word (protein molecule) randomly is 10^{-325} (i.e., almost zero). For this reason, protein synthesis is an order-generating process since the cell rules out almost the whole space of possible proteins and concentrates on a very small portion of some thousands or millions of different proteins.

To describe how DNA is involved in protein synthesis, we need to know something first about the central dogma of molecular biology and the genetic code. The central dogma of molecular biology means that normally DNA does not serve as the direct template for the formation of proteins. Rather, the information for making proteins is first transcribed (copied) into RNA, and then the ribo-nucleotide sequence of RNA is translated into the amino acid sequence of proteins. Accordingly, there are three major classes of RNA: the messenger RNA (m-RNA), transfer RNA (t-RNA), and ribosomal RNA (r-RNA). Messenger RNA carries the coded information for amino acid sequences of proteins. Transfer RNAs form adaptor molecules with the capacity to become specifically attached to amino acids and to complementary base pair with m-RNA molecules. Ribosomal RNAs combine with proteins to form ribosomes; the latter serve as "jigs" for supporting m-RNA and t-RNA during protein synthesis.

Based on mathematical calculations, the simplest genetic code is a triplet; the smallest number of deoxyribonucleotides that can code for an amino acid is three. This conclusion stems from the fact that there are only four bases in DNA (A,

T, C, G) but twenty amino acids commonly found in proteins. Thus a singlet code would not do, because if each base corresponded to an amino acid, there is information for only four amino acids. A doublet code for only $4^2 = 16$ amino acids. A triplet code, however, has $4^3 = 64$ possible combinations, much more than required for the twenty amino acids. Among the 64 possible triplets, two to as many as six different combinations may translate as the same amino acid and three triplets do not specify any amino acid but apparently function as "punctuation" in the set of instructions indicating the end of the message.

Triplets in messenger RNA are called codons. These codons are usually indicated in the literature of molecular genetics as three-letter words, the letters being the initial letters of the four nitrogenous bases occur in RNA. Thus U U U, the first codon discovered and the one that broke the code, is a triplet in which uracil is the base in three successive nucleotides. This is translated as the amino acid phenylalanine. The correlation between codons and amino acids is referred to as the genetic code, and this code is assumed to be universal among organisms.

As mentioned before, the DNA performs two basic functions: self-replication, which secures hereditary information transmission from one generation to the next, and protein synthesis. DNA structure provides the basis for DNA replication (duplication). The hydrogen bond between the base pairs break and the double helix unwinds; each of the two strands thus separated serves as a template on which a complementary strand is formed. In the presence of DNA—polymerase the new strand is synthesized, nucleotide by nucleotide, of units contacting the appropriate complementary purines and pyrimidines.

DNA involvement in protein synthesis is now established to have the following general features: the DNA of a particular gene (group of nucleotides) manufactures a molecule of m-RNA. The m-RNA possesses a complement of the order of nucleotides in DNA (except that there is uracile in all place where thymine exist in DNA). The sequence of nucleotides in a particular m-RNA reflects the sequence in a segment of DNA. The information in the gene is thus transcribed into an m-RNA molecule, which can leave the nucleus and move the site of protein synthesis.

The m-RNA that moves from the nucleus into the cytoplasm becomes associated with a ribosome, in some cases with a group of ribosomes called a polyribosome. Here the m-RNA serves as a template that determines the linear sequence of amino acids in the particular polypeptide being synthesized. The amino acids are brought to the site of this synthesis and one by one attached to molecules of transfer RNA (t-RNA). T-RNA, which has lower molecular weight than m-RNA, exist in at least as many different forms as there are different amino acids, each kind serving to transfer one particular amino acid to the forming polypeptide chain. The peptide linkages between successive amino acids are brought about by specific enzymes. After each amino acid is peptide bonded to the carboxyl terminus of the peptide chain, the t-RNA molecule is released from the complex and recycled through the activation process. Upon completion of the addition of amino acids to the growing chain, the mRNA-ribosome complex is dismantled; the fully formed polypeptide chain is released, and the ribosome is reused through the attachment to another m-RNA molecule in order to produce another protein.

In addition to this peculiar process of bioinformation generation, which is unparallel with regard to inanimate systems, living systems are also capable of transmitting this bioinformation through successive generations (heredity) and maximizing it phylogenetically (evolution).

Heredity:

Gregor Mendel (1822-1884), an Austrian monk who studied mathematics and science, was interested in the possible outcome of types of flowers and fruits that would result from crossbreeding two plants. In particular, the connection between the color of a pea flower and the type of seed that same plant produced inspired Mendel to begin experimenting with garden peas in 1860. Mendel's careful use of scientific methods resulted in the first recorded study of how traits pass from one generation to the next. After eight years, Mendel presented his following results with pea plants to scientists:

1. Heredity material is particulate (rather than fluid).

2. Each individual has two particles (factors) for each trait.

3. Parents make equal contributions to their offspring.

4. Only one pair of each hereditary factors is carried by a gamete (now called the principle of allelic segregation).

5. One factor may produce its effect when in single dose (dominance) and the other member of the pair when in double dose (recessiveness).

6. Each factor is distributed to a gamete independently of other factor pairs (now called the principle of independent assortment).

7. The results of crossing (mating) individuals of known heredity can be statistically predicted on a population basis to occur in ratios such as 3:1, 1:1, and 9:3:3:1.

Before Mendel, scientists mostly relied on observation and description, often studying many traits at one time. Mendel was the first to trace one trait through several generations. He was also the first to use the mathematics of probability to explain heredity. The use of math in plant science was a new concept and not widely accepted then. Moreover, Mendel's principles lacked a mechanical basis for the regular distribution of his hypothetical "factors." Thus Mendel's work was forgotten for a long time.

Due to the developments made in the field of cytology from 1875-1900, the nucleus and chromosomes were discovered. The chromosomes were seen to split prior to cell division, and the processes of mitosis and meiosis were gradually worked out. Consequently, a mechanistic process became available as a basis for the regular distribution postulated for Mendelian factors. In 1903, the word gene was coined to replace Mendel's factor. Based on these developments, genes on chromosomes control an organism's form and function. The different forms of a trait that a gene may have are called alleles. When a pair of chromosomes separates during meiosis, alleles for each trait also separate into different sex cells. As a result, every sex cell has one allele for each trait. The allele in one sex cell may control one form of the trait, such as having blue eyes. The allele in the other sex cell may control a different form of the trait—not having blue eyes. The study of how traits are inherited through the interactions of alleles is the science of genetics. Heredity, the passing of traits from parent to offspring, has the following principles:

1. Traits are controlled by alleles on chromosomes.

2. An allele's effect is dominant or recessive.

3. When a pair of chromosomes separates during meiosis, the different alleles for a trait move into separate sex cells.

G. H. Hardy and W. Weinberg concerned themselves with the theoretical results of random mating in a large population with natural selection in operation. They independently discovered that the frequency of a given gene is not expected to change from one generation to the next. This principle, now called Hardy-Weinberg law, defines static-nonevolving situation and has become the corner stone of population genetics. The Hardy-Weinberg law proved to have great importance to evolutionary theory, since the evolutionary potential of species resides in the store of genetic variability in its gene pool.

Evolution

According to Stansfield (1977) the concept of evolution may be traced back to the Greeks, when Empedocles (504-433 BC) not only asserted that change is the ultimate reality but also noted the following:

- Higher forms of life gradually evolve.

- Plants evolved before animals.

- Better-adapted forms of life tend to replace less-adapted forms.

Stansfield also indicated that Aristotle (384-322 BC) proposed a "ladder of nature," a taxonomic scheme whereby nature is viewed as a continuum from inanimate matter, from plants to lower animals to higher animals, ending with mankind. He asserted that this progression in complexity was under the guidance of divine intelligence. Stansfield (1977) criticized and indicated that Aristotle's concept of biology was both vitalistic and teleological and that his ladder was a device for classification rather than a theory of explaining the origins of organic diversity.

During the Middle Ages and up to the eighteenth century, the ideas of special creation and immutability of species, which were a literal interpretation of the Bible, dominated the scene. To the extent that even Carolus Linnaeus (1707-1778) and George Cuvier (1769-1832) were discouraged to speculate on evolution although their findings in taxonomy and paleontology suggest strongly the evolution of biological systems. For example, it was clear that fossils represent species and genera not found among living creatures all fitted neatly into one or another of the known phyla and so made an integral part of the scheme of life. Furthermore, the deeper the stratum in which the fossil was to be found, and therefore the older the fossil, the simpler and less highly developed it seemed. Not only that, but fossils sometimes represented intermediate forms connecting two groups of creatures, which as far living forms were concerned, seemed entirely separate.

This was the main obstacle that hindered the efforts to explain the varieties of life: If animals had evolved from one to another, what had caused them to do so? The first to attempt an explanation was French naturalist Jean Bapteste de Lamark. Contrary to Cuvier, who regarded species as immutable and fixed since their creation, Lamark believed that species were not static but were derived from preexisting species. He systematically placed organisms in a series from the simplest to the most complex. In 1802, Lamark published the first mechanistic theory of evolution, called the theory of use and disuse or the inheritance of acquired characteristics. According to this theory:

i. Structural variations are acquired because of need.

ii. Use of a structure increases its size; failure to use it results in its atrophy or disappearance, which is the theory of use and disuse.

iii. These variations, which are now referred to as acquired characteristics, are inherited. Thus the progeny have the advantage of the favorable adaptations acquired by their parents.

Although in general there is some truth in the principle of use and disuse, there is no evidence for the first and third phases of this theory. The assumption that acquired characteristics are inherited was challenged by August Friedrich Leopold Weismann (1834-1914) as early as 1892, when his principal work was published. Weismann drew the distinction between the germ plasm and the soma. The germ plasm has its seat in the sex cells. The soma is the rest of the body. The germ plasm is potentially immortal; indeed, every sex cell is able, under favorable conditions, to give rise to a new individual with another crop of sex cells. The soma is mortal; it is the body that houses the sex cells and is cast off in every generation owing to death. Weismann's famous experiment consisted in cutting off the tails of newborn mice in a series of successive generations. The tails were no shorter in the progeny of experimental mice; consequently, he concluded that evolutionary changes could not occur unless they were incorporated in the germ plasm.

The Theory of Darwin:

The two major aspects of evolution are the origin of variation (that is, the appearance of differences between organisms) and the origin of species. Darwin did not understand the origin of inherited variation or the mechanisms of their transmission. Nevertheless, he went on to explain the origin of species by natural selection. The logic by which he came to his natural selection theory can be summarized by the following:

1. Organisms are prodigal in their production of offspring, far too many being produced to survive.

These progeny are not alike; they show much fortuitous variation.

2. This prodigal production results in competition or struggle for existence.

3. Competition leads to natural selection of the most fit through the deaths of those less fit to survive.

4. The progeny of the organisms most fit to survive inherit the characteristics of their parents, namely those characteristics that have made their parents most fit.

This is Darwin's selection theory, which, contrary to Lamark's theory, was able to assimilate the subsequent development in biological research and to metamorphose in what is known, in the present times, as neo-Darwinism or the Modern synthesis.

Neo-Darwinism is a more developed version of Darwin's theory of evolution by natural selection, which became dominant and reached a certain standard of completion between 1930 and 1950. It has also been called the synthetic theory because it is a synthesis of Darwin, Mendel, mutations, and modern statistical methods. Stebbins (1977) indicated that this fusion of originally separate ideas is a cooperative achievement of many workers, among them S. S. Chetverikov, R. A. Fisher, J. B. S. Haldane, S. Wright, Th. Dobzhansky, S. J. Huxley, E. Mayer, G. G. Simpson and G. L. Stebbins.

These workers have been able to substantiate the following main assumptions that characterize the modern synthesis:

a. Genetic changes underlie evolution.

b. Mutations (in the modern general sense) are the ultimate source of evolution. These may be small (point mutations) or large (chromosomal), but the former are more likely to lead to advantageous changes than the latter and hence have been more important in evolution. However, the main point is that they are random in the sense of not being directed.

c. The central dogma and Weismann's principle apply.

d. Evolution is defined in terms of change in gene frequencies.

e. These changes may occur by mutation, movement of gene into and out of populations, random drift, and natural selection, but the latter is by far the most important cause.

These principles are sufficient to explain the diversity and adaptation of organisms on earth. This powerful explanatory and predictive theory has become the central organizing principle of modern biology, directing research and providing a unifying explanation for the history and diversity of life on Earth. Evolution is applied and studied in fields as diverse as agriculture, anthropology, conservation biology, ecology, medicine, paleontology, philosophy, and psychology.

1.2—DEFINITION OF LIFE

Are the above-mentioned characteristics or hallmarks (i.e., carbon-based chemistry, metabolism, information system, heredity, and evolution) sufficient to define life phenomenon? If defining life is revealing its essence or nature, the answer is no, because revealing the nature of life implies explanation rather than just description. The purpose of description is usually to answer the question what, by mentioning the relevant attributes of the object, whereas explanation necessitates answering the question of why the object is such as it is. Moreover, different scientists used to propose different hallmarks or criteria. For example, Ganti's (2003) hallmarks fall into two categories: real (or absolute) and potential. Real-life criteria specify the necessary and sufficient conditions for life in an individual living organism. These are the proposed real-life criteria:

1) Holism: An organism is an individual entity that cannot be subdivided without losing its essential properties. An organism cannot remain alive if its parts are separated and no longer interact.

2) Metabolism: An individual organism takes in material and energy from its local environment and chemically transforms them. Seeds are dormant and therefore lack an active metabolism, but they can become alive if conditions reactivate their metabolism.

3) Inherent stability: An organism maintains homeostatic internal processes while living in a

changing environment. By changing and adapting to a dynamic external environment, an organism preserves its overall structure and organization.

4) Active information-carrying systems: A living system must store information that is used in its development and functioning. Because the information can be copied, children inherit this information through reproduction. Mistakes in information transfer can "mutate" this information, and natural selection can sift through the resulting genetic variance.

5) Flexible control: Processes in an organism are regulated and controlled to promote the organism's continued existence and flourishing. This control involves an adaptive flexibility and can often improve with experience.

In contrast to these real criteria, Ganti also proposed potential life criteria. A living individual organism can fail to possess life's potential criteria. The defining feature of potential life criteria is that if enough organisms exhibit them, then life can populate a planet and sustain itself. Ganti proposed three:

1) Growth and reproduction. Old animals and sterile animals and plants are all living, but none can reproduce. Therefore, the capacity to reproduce is neither necessary nor sufficient for being a living organism. But due to the mortality of individual organisms, a population can survive and flourish only if some organisms in the population reproduce. In

this sense, growth and reproduction are what Ganti calls potential rather than real-life criterion.

2) Evolvability. "A living system must have the capacity for hereditary change and, furthermore, for evolution, i.e., the property of producing increasingly complex and differentiated forms over a very long series of successive generations" (Ganti 2003). Since what evolves over time are not individual organisms but populations of them, we should instead say that living systems can be members of a population and have the capacity to evolve.

3) Mortality. Living systems are mortal. This is true even of clonally asexual organisms, for death can afflict both individual organisms as well as the whole clone. Systems that could never live cannot die, so death is the property of things that were alive.

Bedau (2006) indicates that Ganti's life characteristics or hallmarks and other lists of life's hallmarks always reflect and express some preconceptions about life. This might seem to beg the question of what life is. He emphasizes that any nonarbitrary list of life's hallmarks was presumably constructed by someone using some criterion to rule examples in or out. But where did this criterion come from, and what assures us that it is correct? Why should we be confident that any hallmarks that fit it reveal the true nature of life? With regard to these questions, Bedau concludes that lists of life's hallmarks cannot be the final word on what is life; as we learn more about life, our preconceptions change, evolve, and mature. Then the same should be expected of our lists of life's hallmarks. Bedau asserts that a more comprehensive definition of life that reveals its essence must account, in addition to hallmarks, for borderline cases and puzzles. These are:

1) He characterizes viruses as familiar examples of borderline cases that self-replicate and spread even though they have no independent metabolism.

2) Dormant seeds or spores are another kind of borderline case, the most extreme version of which might be bacteria or insects that are frozen.

3) Populations of microscopic clay crystallites growing and proliferating are another kind of borderline example, especially because they can in appropriate circumstances undergo natural selection (Bedau 1991).

4) Another kind of borderline case consists of soft artificial life creations like Tierra. Tierra is software that creates a spontaneously evolving population of computer programs that reproduce, mutate, and evolve in computer memory. Tierra's inventor thinks that Tierra is literally alive (Ray 1992). This would radically violate the ordinary concept of life that most of us have.

5) One final category of borderline cases consists of complex adaptive systems found in nature, such as financial markets or the World Wide Web. These exhibit many of the hallmarks of life, and some think that the simplest and most unified explanation of the

entire range of phenomena of life is to consider these natural complex adaptive systems to be literally alive (Bedau 1996, 1998).

Puzzles about Life

According to Bedau, any account of life should explain the origin of the following puzzles; more important, it should resolve the puzzles:

- Origins: How does life or biology arise from nonlife or pure chemistry? What is the difference between a system that is undergoing merely chemical evolution, in which chemical reactions are continually changing the concentrations of chemical species, and a system that contains life?

- Emergence: B properties are said to emerge from A properties when the B properties both depend on, and are autonomous from, the A properties. Different kinds of dependence and autonomy generate different grades of emergence (Bedau 2003). One is the strong emergence involved in irreducible top-down causation.

- Synchronic: It concerns what properties exist at a moment. Life is the paradigm case of a dynamic form of "weak" emergence, one that concerns macro properties that are unpredictable or underivable except by observing the process by which they are generated, or by observing a simulation of it (Bedau 1997, 2003).

- Hierarchy: Various kinds of structural hierarchies characterize life. Each organism has a hierarchical internal organization, and the relative complexity of organizations of different kinds of organisms form another hierarchy. Bedau asks why life tends to generate and encompass such hierarchies. This question applies both to the hierarchy in complexity that spans all organisms together as well as to the organizational hierarchy found within each individual living organism.

- Continuum: Can things be more or less alive? Is life a black-or-white Boolean property or a continuum property with many shades of gray? Bedau indicates that although common sense leans toward the Boolean view—a rabbit is alive and a rock isn't, for example—in fact, there are borderline cases like viruses that are unable to replicate without a host.

- Strong artificial life: Artificial life software and hardware raise the question of whether our computer creations could ever literally be alive.

- Mind: Another puzzle is whether there is any intrinsic connection between life and the human mind. Plants, bacteria, insects, and mammals, for example, exhibit some kind of intelligent behavior such as sensitivity to the environment, various ways in which this environmental sensitivity affects their behavior, and various forms of inter-organism communication. So it is natural to ask whether life and mind have some deep connection.

To develop a unified theory of life that explains life hallmarks, borderlines, and puzzles, it is reasonable initially to focus on revealing the nature of life: whether it is a purely physical phenomenon or whether it is independent of physics. This is the dilemma of reductionism and antireductionism.

1.3—REDUCTIONISM AND ANTIREDUCTIONISM DILEMMA

As we said earlier, living systems are problematic in the sense that although they are composed of the same elements found among inanimate systems and do not violate the laws governing the physicochemical transformations and reactions of these elements, they grow, develop, and evolve, and thus generate bioinformation or biocomplexity in a peculiar manner unattainable to inanimate systems. So how can we explain their peculiar behavior? Are they complicated machines, subject to the same physicochemical laws governing their elementary inanimate constituents and transformations? Are they autonomous systems that obey new laws that are independent of physics? Are they chaotic, haphazard, indeterminate systems? In short, what is the nature of life?

The problem of the nature of life, put in this manner, is sometimes called the "problem of reductionism" or the "question of reduction," which has a long history of controversy and heated debates filled with emotions. Reductionism is underpinned by the doctrine of causal closure or completeness of physics. The stronger formulations of causal closure assert this: "No physical event has a cause outside the physical domain" (Kim 1998) or

"Physical effects have only physical causes" (Vincente 2006). That is, the stronger formulations assert that for physical events, causes other than physical causes do not exist. In this perspective, living systems having physical effects entail that biology is reducible to ordinary physics.

Ayala (1977) analyzed this important question and distinguished between the ontological, methodological, and epistemological aspects of reductionism. Ontological reductionism is concerned with questions of structural or constitutive nature (e.g., whether or not physicochemical entities and processes underlie all living phenomena). Are organisms constituted of the same components as these making up inorganic matter? Or do organisms consist of other entities besides molecules and atoms? Ontological reductionism claims that organisms are exhaustively composed of inanimate parts. It also implies that the laws of physics and chemistry fully apply to all biological processes at the level of atoms and molecules.

In the ontological domain, Ayala noted that reductionism versus antireductionism in its extreme form resolves into mechanism versus vitalism. The mechanists' position is that organisms are ultimately made up of the same atoms that make up inanimate matter and nothing more. Vitalists argue that organisms are made up of not only material components (atoms and molecules and aggregation of them) but also of some nonmaterial entity, variously called entelechy, vital force, vital energy, and the like. Aristotle (384-322 BC), the great Greek philosopher, is sometimes said to be the first proponent of vitalism. The modern controversy over mechanism versus vitalism dates from the seventeenth century, when Rene' Descartes (1596-1650) proposed that animals are nothing other than complex machines. Early in the twentieth century, vitalism was refined by such philosophers as Henri Bergson (1859-1941) and by some biologists, notably Hans Driesch (1867-1941). At present, vitalism has no distinguished proponents among biologists because it does not meet the requirements of a scientific hypothesis. Vitalism is not a hypothesis subject to the possibility of empirical falsification and therefore leads to no fruitful observations or experiments. However, recently a new form of vitalism developed, called intelligent design movement, capitalizing mainly on the difficulties facing neo-Darwinism.

According to Ayala, as mentioned above, methodological reductionism concerns the strategy of research and the acquisition of knowledge, the approaches to be followed in the investigation of living systems. The general question is whether biological problems should always be investigated by studying underlying (ultimately physical) processes or whether they should also be studied at higher levels of organization, such as the cell, the organism, the population, and the community. Methodological reductionism claims that living phenomena are best studied at lower levels of complexity, ultimately at the level of atoms and molecules. Methodological reduction has its counterpart in methodological compositionism according to which organisms and group of organisms should be studied as wholes. Having discussed both approaches, Ayala concluded that methodological reductionism and compositionism are complementary; often the best strategy of biological research is an alteration between analysis and synthesis.

The third type of reductionism discussed by Ayala (1977) concerns issues that may be called epistemological, theoretical, or explanatory. The fundamental issue here is whether the theories and laws of biology can be derived from the laws and theories of physics and chemistry. Epistemological reductionism of one branch of science to another takes place when the theories or experimental laws of a branch of science, called the secondary science, are shown to be special cases of the theories and laws formulated in some other branch of science, called the primary science. The integration of diverse scientific theories and laws into more comprehensive ones simplifies science and extends the explanatory power of scientific principles, and this conforms to the goals of science

Nagel (1961) has formulated the two conditions that are necessary and sufficient for the reduction of one theory or branch of science to another. When stated with special reference to biology and physico-chemistry, they are as follows:

a. The condition of connectability: All terms in a biological law that do not belong to the primary science (such as cell, mitosis, or heredity) must be connected with expressions constructed out of the theoretical vocabulary of physics and chemistry (out of energy and the like).

b. The condition of derivability: Every biological law, whether theoretical or experimental, must be logically derivable from a class of statements belonging to physics and chemistry.

On applying these conditions to biology, Nagel (1961) and Ayala (1977) reached the following conclusion: "The impressive successes of molecular biology during recent decades have moved some people to claim that the only worthy and truly scientific biological investigations are those leading to the explanation of biological phenomena in terms of their underlying physico-chemical components and processes. Nevertheless, epistemological reduction of biology to the physical sciences is not possible at present. In the current stage of scientific development, a great many biological terms such as 'organ,' 'species,' 'consciousness,' 'mating propensity,' 'fitness,' 'competition,' and 'predator' cannot be defined adequately in physico-chemical terms. Nor is there any class of statements and hypothesis in physics and chemistry from which every biological law can be derived logically. Therefore, neither the condition of connectability nor the condition of derivability—the two necessary conditions for epistemological reduction—can be satisfied." It is appropriate to emphasize that although half a century has passed, and despite the enormous quantification of biology, the conditions proposed by Nagel are still far from being realized. Moreover, H. H. Pattee (1968) asserts that if living matter is the same as nonliving matter with respect to description by physical laws, then this does not answer the obvious question of why living matter is so conspicuously different from nonliving matter.

On the other hand, antireductionists claim that life is irreducible to physics and that life is an emergent property of matter at high complexity. It follows that the obvious biological properties characteristic of living systems are due to their enormous organization. The antireductionists, with such a claim, face two difficulties: Either they regard that there is an independent set of laws concerning biological systems. In this case, they have to consider Schrödinger (1944), Elsasser (1981), and Pattee's (1971) emphasis that there should be no new set of laws independent of quantum mechanics. Or they may regard biological phenomenon and in particular evolution a random process that does not accord with the growing concern that beneficial mutations are not random (Perez 2010; Elsheikh 2010), in addition to numerous scientists (which we will cover in the next chapter) who object to the randomness of biological evolution based on statistical improbability. It follows that both reductionism and antireductionism face a deadlock.

CHAPTER 2

THE PROBLEM WITH EVOLUTION

2.1—INADEQUACIES OF NEO-DARWINISM:

As we said earlier, the modern theory of evolution, neo-Darwinism, emerged in the 1930s and 1940s as a synthesis of genetic knowledge and the Darwinian concept of natural selection. As a result, evolution is regarded as a process generated by the operation of natural selection on random mutational changes. Neo-Darwinism has gradually pervaded all biological disciplines and contributed to their development and enrichment. At the same time, neo-Darwinism has been expanded by the contribution from other biological disciplines, such as zoology, botany, anthropology, paleontology, physiology, microbiology, biochemistry, and experimental and mathematical population biology. This is why neo Darwinism is often called the synthetic theory precisely because it integrates the contributions of so many fields of knowledge. For decades, neo-Darwinism stood virtually unchallenged as the basis of our understanding of the organic world.

Today the picture is entirely different. The difficulties concerning the recognition of the nature of life and the lack of a general unified theory of biology naturally obscure the understanding of biological evolution. More and more workers are showing signs of dissatisfaction with the synthetic theory. Some are attacking the concept of randomness, charging that if natural selection has to choose from the astronomically large number of the alternative systems by means of the mechanisms described in modern evolution theory, the chances of producing a creature like us is virtually zero (Bertalanffy 1954; Wadington 1968; Gould 1982; Yockey 1992). Eden, who was especially concerned about the elements of randomness, contended, "No currently existing language can tolerate random changes in the symbol sequences which express its sentences. Meaning is almost invariably destroyed. Any changes must be syntactically lawful ones" (Eden 1966).

Bohr (1933) had the view, which we adopt, that the distinction between living and nonliving systems was fundamental and is a manifestation of his principle of complementarity. However, it was not clear at that time what characterizes such living and nonliving systems' complementarity. Schrödinger emphasized that the principle that generates order from order is not alien to physics. He iterated, "It is, in my opinion, nothing else than the principle of quantum theory over again" (Schrödinger 1944). McClendon (1980) compared biological evolution with chemical evolution of isotopes. From the comparison, he concluded that the forces that drive biological evolution are intrinsic property of matter. In other words, the evolution of novel, more complex organisms from lower ones precedes adaptation and selection. Kauffman (1995) had a similar view. He suggests that selection acts not only on random variations but also on emergent patterns of order that self-organize via the laws of nature.

Whyte (1965) suggested that in addition to Darwinian selection there should be an internal selection of mutants at the molecular, chromosomal, and cellular levels, in accordance with their compatibility with internal coordination of the system. Waddington (1968) tried to show that evolution does not depend on random search. He emphasized that what occurs randomly are the mutations on the genome level; however, the output of these changes on the phenotype is not random (i.e., there are certain operators that map the space of genotypes into a "fitness space"). Dawkins (1986) proposed what he called cumulative selection as an alternative to what he called single-step selection.

Perez (2010) demonstrated that an evolutionary matrix governs the structure of DNA so that beneficial mutations cannot be random. If evolution is not a purely random process, then it could be either a miracle or an expression of natural law, i.e., an expression of a life-organizing principle. Elsheikh (2010) showed that biological evolution is subject to a life function (life-organizing principle), which is a generalized Schrödinger type of system, with vitality (a measure of bioinformation) as path variable. It follows that beneficial mutational changes is a process through which the life-organizing principle undertakes negative damping, leading to the amplification of bioinformation oscillations and the consequent increase of total vitality as an evolutionary goal function or target criterion.

Accordingly, Neo-Darwinism, in absence of a unified theory of biology that identifies the nature of life, is plausible not to have adequate account for the following dichotomies: randomness versus determinism as well as ontogeny and phylogeny and gradualism versus punctualism. It is appropriate to note that by revealing the inadequacies and limitations of neo-Darwinism theory, we are not aiming at discrediting the validity of the theory; rather, we would like to emphasize as Eldredge (1985) did: "Throughout the book I will conclude that the synthesis is not so much incorrect as complete."

Randomness and Determinism:

A widely publicized debate between a group of mathematicians and biologists regarding the problem of randomness took place in April 1966 at Wister Institute of Anatomy and Biology in Philadelphia. The mathematicians charged that if natural selection has to choose from the astronomically large number of the alternative systems by means of the mechanisms described in current evolution theory, the chances of producing a creature like us is virtually zero. Eden (1966) who was especially concerned about the elements of randomness, referring to genetic code as a language, contends that no currently existing language can tolerate random changes in the symbol sequences that express its sentences. Meaning is almost invariably destroyed. Any changes must be syntactically lawful ones.

Therefore, not all attempts to simulate the evolutionary process on computers have been successful, and the mathematicians concluded that current evolutionary theory is inadequate. It has to supply the programmer with a correct set of rules for "genetic grammaticality," which has a deterministic explanation rather than owing the observed stability of biological systems to selection pressure acting on random variations. Additional evidence for the nonrandom variation and for the view that evolution is self-limiting is emphasized by the discovery that amino acids are self-sequencing (Fox 1984). Fox then concluded that the fact that amino acids are self-sequencing or self-ordering, which is the essence of an evolutionary theory that takes us beyond and out of neo-Darwinism.

Probably based on Eden's (1966) objection to random search and on his assertion that: "No currently existing language can tolerate random changes in the symbol sequences which express its sentences. Meaning is almost invariably destroyed. Any changes must be syntactically lawful ones", Dawkins (1986) proposed what he called cumulative selection. He clarified the concept by drawing attention to the fact that given a sentence of twenty-eight units, the probability of a monkey typing it right away (what he called single-step selection) is negligible. However, if whenever a letter falls in its proper place, if it is preserved so that the next random changes act on the remaining letters; then the chance for writing the sentence in this manner, which Dawkins calls cumulative selection, is very much improved. It is obvious that the process of generating a meaningful sentence in Dawkins's above-mentioned example is guided by English language grammar. However, since such grammar or set of rules to guide the evolutionary process has not yet been discovered, Dawkins made some reservation as to the significance of his example.

Thus it is notable that the trend that evolution is nonrandom and deterministic is gaining momentum. However, this necessitates that the deterministic view should be capable of resolving the basic problems that neo-Darwinism fails to solve according to its conception of evolution as a random process, namely the problem of the direction or goal function that characterizes the evolutionary process. Regarding the former Kammshilov (1976), he noted that a distinction is not always clearly drawn between two notions: the motive forces of the evolutionary process and its adaptive form. He clarifies this distinction with the analogy that the water in a river flows obeying the force of gravity, while the form of the riverbed is determined by the landscape that forms the configuration of its banks. There is no river without banks, but it is not the banks that move water in the river.

Because neo-Darwinism conceives evolution as random process, it is helpless to find out any direction for evolution or measure for biological progress. Accordingly, Dobzhansky (1955) suggested more than one criterion, (i.e., more than one direction for evolutionary progress), namely increases of complexity, increase of biomass, and improvement of homeostatic adjustments. Moreover, referring to Simpson, Ayala (1977) also emphasized that there is no standard according to which uniform progress can be said to have occurred in the evolution of life. Changes of direction, slackening, or reversals have occurred in all evolutionary lineages, no matter what feature is considered.

Ontogeny and Phylogeny:

Phylogenetic or evolutionary changes are usually described in terms of the differences between successive adults. However, the differences between those adults were the consequence or differences between the paths of development that gave rise to them (that is, ontogenetic changes). This is why Whyte (1965) asserted, as some embryologists held, that ontogeny is theoretically primary to phylogeny; consequently, the synthetic theory cannot be regarded as definitive until it has been combined with a theory of ontogeny. Lovtrup (1984) also emphasized the intimate association between ontogeny and phylogeny. He noted that ontogeny recapitulates the historical as well as the creative aspects of phylogeny. But he did not claim that the resolution is such that it is possible to make phylogenetic classifications on the sole basis of embryological studies.

If ontogeny and phylogeny are so obviously and necessarily related, what impedes neo-Darwinism from incorporating this relation? To answer this question, we shall refer to Saunder (1984), who noted that neo-Darwinism contains no theory of the origin of variations. It simply assumes that while they must be small, they are to all intents and purposes random. A character is seemed to be adequately explained if we can determine, or at least postulate, its selective advantage and evolutionary history. Accordingly, what puts phylogeny and ontogeny at disparity is an extrapolation of Weimann's doctrine of the independence of the germ plasm, according to which the only variations that matter in evolution are random genetic mutations. Now, Saunder argued that if we suppose that the source of all heritable variation is random mutation, it still does not follow that the variations in the phenotype are random. They may be triggered by mutations, but the forms they take are dependent on the properties of the epigenetic system. Thus Saunders emphasized that the key to an understanding of the origin of variation, and hence to an understanding of evolution itself, must lie in the study of development.

But what do we mean by an epigenetic system or field? And how can it account for a possible link between phylogeny and ontogeny or modify Weismann's doctrine or the central dogma to account for such a link? Waddington (1968) explained that he introduced the word "epigenetic" in 1947 as a suitable name for the branch of biology that studies the causal interactions between genes and their products that bring the phenotype into being (i.e., the causal study of development). He distinguished between two aspects of epigenetics: changes in cellular composition (cellular differentiation) and changes in geometrical form (morphogenesis). In his view, these changes are canalized in the sense that epigenetic trajectories normally show some resistance to being changed. He introduced developmental as well as genetic evidence, revealing the self-regulatory nature of the epigenetic field. At the same time, he made clear that the self-regulating aspect of the epigenetic trajectories is a phenomenon of a more general nature than homeostatic in that the thing that is being held constant is not a single parameter but is a time-extended course of change (trajectory). He regarded this situation as one of homeorhesis, stabilized flow rather than stabilized state.

Later on, when Waddington (1968) tried to show that evolution does not depend on random search, he proposed that we start with a population of genotypes in a multidimensional "genotype space." These are mapped, through a multidimensional space of epigenetic operators (in which operators arising from the environment are included), into an also multidimensional space of phenotypes. This is then mapped by some complex function into an essentially one-dimensional fitness space in which the only variable is the coefficient of fitness, (i.e., the number of offspring produced).

To overcome the limitations of the genotype-phynotype duality, which obstructs a proper conceptualization of the relation between development and evolution, Goodwin

(1984) proposed a field theory of the generative process: morphogenetic field. He summarized his idea in the following formula: particulars + universals = specific form. The generative field equations embody the universals and the parameters containing the particulars. The latter includes the two sources of particular influences on organisms, those that are inherited and those that come from environment. The molecular elements that make up the former, the generative field, are either gene products or are substances whose concentrations depend upon the regulatory activities of gene products. He proposed that the organizational properties, insofar as they relate to the generation and regeneration of organism morphology, are to be at least partly understood in terms of the principles of continuum mechanics as expressed in viscoelastic field equations.

Thus, according to Goodwin (1984), the attempt to understand biological process in historical terms, the objective of the modern synthesis based upon the evolutionary paradigm, has failed because of the impossibility of explaining observed biological order and regularity in terms of contingencies and differences, or accidental variation. Then he concluded: "But the momentum of biological research in this century has been such that many of the ingredients for a rewriting of Weismann's fruitful but limited scheme, uniting development and evolution within a logically ordered generative process which includes not only hereditary particulars but also biological universals appropriate to different levels of organization in the organism, seems now to be a distinct and existing possibility. Such a rewriting, if it can be realized, will transform biology from an historical to an exact science" Goodwin (1984).

Gradualism Versus Punctualism:

From Darwin onward, evolutionists have revealed that if we arrange all our available fossils in chronological logical order, they do not form a smooth sequence of gradual change. Most evolutionists following Darwin have assumed that this is mainly because the fossil record is imperfect. However, Eldredge and Gould (1972) suggest that actually the fossil record may not be as imperfect as we thought. Maybe the gaps are a true reflection of what really happened, rather than being an imperfect fossil record. They suggested that evolution may in some sense go in sudden bursts, punctuating long periods of stasis. Their theory of punctuated equilibrium postulates that evolution within species differs from that which establishes new species during lineage splitting. Relatively more phenotypic change is thought to be associated with punctuation and less net change to occur within species (i.e., during stasis). Evolution within species is known as microevolution, and this consists largely of all the substitutions prompted by natural selection: mutation, genetic drift, operating at the level of the individual organism. Thus microevolutionary changes are gradual and rarely, if ever, yield any substantial morphological change. On the other hand most morphological change is associated with the origin of new species, genera and higher taxa (i.e., macroevolution), whereby the relevant unit of study is the species rather than the individual organism.

Now an important question facing neo-Darwinism is whether the mechanisms underlying microevolution can be extrapolated to explain macroevolution, namely to explain punctuated equilibrium. Ayala (1983) analyzed this problem to see whether macroevolution is reducible or autonomous relative to microevolution. The issue of whether the mechanisms' underlying microevolution can be extrapolated to macroevolution involves, in his view, at least three separate issues to consider:

1. Whether microevolutionary processes operate (and have operated in the past) throughout the organisms that make up the taxa in which macroevolutionary phenomena are observed

2. Whether the microevolutionary processes identified by population genetics (mutation, random drift, natural selection) are sufficient to account for the morphological changes and other macroevolutionary phenomena observed in higher taxa or whether additional microevolutionary processes need to be postulated

3. Whether theories concerning evolutionary trends and other macroevolutionary patterns can be derived from knowledge of the microevolutionary process

Answering the first issue, Ayala says that it seems unlikely that any paleontologist or macroevolutionist would claim that mutation, drift, natural selection, and other microevolutionary processes do not apply to the organisms and populations that make up the higher taxa studies in macroevolution. Regarding the second question, he also argued in favor of the view that the known processes of microevolution can account for macroevolutionary change, even when this occurs according to the punctualist model (that is, at fast rates concentrated on geological brief time intervals). He also emphasized that stasis does not necessarily need to postulate new processes yet unknown to population genetics in order to account for the long persistence of lineage. He asserted that the weight of the evidence favors stabilizing selection as the primary process responsible for morphological stasis of lineages through geological time. Stabilizing selection occurs when the environment is rather uniform in space and time, and it is advantageous for the population to limit the range of its variability so that only or mostly phenotypes tested by natural selection are produced (Dobzhansky 1977).

When coming to the third question—Can macroevolutionary theory be derived from microevolutionary knowledge?—Ayala (1983) emphasized this: "The answer can only be no". Because, in his view, If macroevolutionary theories were deducible from micro evolutionary principles, it would be possible to decide between macroevolutionary models simply by examining the logical implications of microevolutionary theory. But the theory of population genetics is compatible with both punctualism and gradualism, and hence it logically entails neither.

In conclusion, Ayala stated that macroevolutionary phenomena are underlain by microevolutionary phenomena and are compatible with microevolutionary processes, but macroevolutionary studies require the formation of autonomous hypotheses and models that must be tested using macroevolutionary evidence. In this epistemologically very important sense, he said that macroevolution is decoupled from microevolution: macroevolution is an autonomous field of evolutionary study. Gould (1983) not only emphasized the decoupling of macroevolution from microevolution but also suspects or does not feel as most neo-Darwinians feel—that stabilizing selection could be a sole explanation of stasis. In his view, selection and drift are often cited as the only important models of evolutionary change, whereas it is time to consider a third major way. "The rarely recognized category (these days) that includes most non-Darwinians proposal of the past century—internally generated changes of the genome sufficiently rapid in occurrence and great in extent to present fundamentally new features as facts accomplish to forces of selection (that may then accept or reject) or drift (that may then fix or eliminate)" Gould (1983).

Saunders (1984) has been more critical toward gradualism. In his view, neo-Darwinists mostly adhere to the traditional view that evolution is a relatively uniform process of slow and continuous change not because of the palentogical evidence but because their theory demands it. Since the theory claims that individual variations are random, then it is exceedingly disastrous for the individual in which it occurred. Saunders believes that much of the force of this argument disappears when we consider development. He noted that biological systems are maintained and develop by nonlinear interactions. The differential equations that describe them typically have solutions with the mathematical property of structural stability, which in this context is equivalent to what biologists call homeorhesis. If a solution is perturbed by fluctuations below a certain threshold level, it will tend to return to its original course and eventually reach the state it would have achieved had there been no perturbation. If the perturbations exceed the threshold, the system will be unable to return to its original course. This may well result in the cessation of development and hence the death of the organism, but it may cause the system to end up in an alternative developmental pathway that will, like the original one, be stable.

Thus Saunders (1984) emphasized the following: "Homeorhesis is a necessary property of an epigenetic system. But a system which possesses this property will also have the capacity for heterohesis, meaning large, organized change. Such a system will tend not to allow small changes to accumulate but will occasionally permit large ones. Consequently we would expect its evolution to be well described by Eldredge and Gould's punctuated equilibria."

2.2—LIMITATIONS OF SELF-ORGANIZATION MODELS

Neo-Darwinism riddled by the dichotomies: determinism versus randomness, phylogeny versus ontogeny, and gradualism versus punctualism, certain authors, particularly those of creationism orientation, regard these inadequacies as decisive refutation of Darwinism, hence call for an intelligent designer (Meyer 2005; Dembski 1998). Other scientists who regard life as a natural phenomenon start developing theories of self-organization, with the underlying assumption that life is the intelligent designer of its own (Nicolis and Prirogine 1997; Bak and Sneppen 1993; Kauffman 1995; Elsheikh2010).

Thus Nicolis and Prigogine (1977) developed an extension of thermodynamics that shows how the second law can allow for the emergence of novel structures and indicate the ways in which order can emerge from chaos. According to them, life is far from an equilibrium nonlinear oscillatory phenomenon.

However, Fath et al (2001) draws attention to the fact that Prigogine's principle applies for open systems near equilibrium, whereas far from equilibrium where living systems operate, the principle does not apply. This is why, in their view, the search for organizing principles that do apply has produced a variety of energy orientors. Moreover, Sewell (2011) asserts that if we define "X-entropy" to be the entropy associated with any diffusing component X (for example, X might be heat)—and since entropy measures disorder, "X-order" would be the negative of X-entropy—a closer look at the equations for entropy change shows that they not only say that the X-order cannot increase in a closed system but also that in an open system, the X-order cannot increase faster than it is imported through the boundary. Thus in his view, the equations for entropy change do not support the spectacular increase in order that has occurred on Earth. It is important to emphasize that biological evolution is a fact and the Earth being an open system is a fact, so if the second law does not account for the enormous growth of order on Earth, this does not mean evolution violates the second law. Rather, it does mean that the second law is not sufficient to account for biological evolution, which opens the door for a possible biological organizing principle.

Based on the same perspective, Kauffman (1995) advanced a self-organizational theory to account for the emergence of novel biocomplexity. Kauffman suggests that selection acts not mainly on random variations but on emergent patterns of order that self-organize via the laws of nature. Kauffman suggests that when the complexity of the system (as represented by buttons and strings) reaches a critical threshold, new modes of organization can arise in the system "for free"—that is, naturally and spontaneously. Kauffman illustrates that for a dynamical system, such as an autocatalytic net, to be orderly, it must exhibit homeostasis—that is, it must be an attractor (1995, 79).

We agree with Kauffman on the self-organizing capacity of living systems; we also agree that this self-organizing capacity is not reducible to physics, if by physics we mean present-day physics (meaning inanimate physics). Nonetheless, Kauffman's model suffers from some inadequacies. First, for example, Meyer (2004) indicates that in the light system, the order that allegedly arises for "for free" actually arises only if the programmer of the model system "tunes" it in such a way as to keep it from either generating an excessively rigid order or developing into chaos. Meyer emphasizes that this necessary tuning involves an intelligent programmer selecting certain parameters and excluding others—that is, inputting information. Second, Kauffman's model systems are not constrained by functional considerations and thus are not analogous to biological systems.

Bak and Sneppen (1993) proposed the theory of self-organized criticality, which refers to the tendency of large dynamical systems to organize themselves into a (poised) state far out of equilibrium with propagating avalanches of activity of all sizes. Despite a significant effort in studying self-organized criticality (SOC) models, Frigg (2003) argued that SOC cannot possibly be a general theory in the same way thermodynamics or Newtonian mechanics are general theories. In his view, SOC models are gross oversimplifications and cannot in any way be considered realistic descriptions of their target systems.

The difficulties facing these self-organization models to address living systems substantiate Elsheikh's (2010) proposed generalized complementarity, according to which a material system (animate or inanimate) does not simultaneously possess matter waves and bioinformation oscillations descriptions. In other words, models applicable to inanimate systems, like the sand pile, do not reveal the dynamical essence of living systems.

2.3—LIMITATIONS OF QUANTUM THEORY

Quantum mechanics is the fundamental physics theory; hence the proposed notion of life-organizing principle should somehow be associated with quantum mechanics. It is true that presently quantum mechanics provides the basis for the shapes and sizes of biological molecules and their chemical affinities. Quantum mechanics also accounts for the strengths of molecular bonds that hold the machinery of life together and permit metabolism. Nonetheless, Davies (2004) emphasized that if quantum mechanics is to play a nontrivial role in biosystems, then some way to sustain quantum coherence at least for biochemically, if not biologically, significant time scales must be found. Without this crucial step, quantum biology is dead. Moreover, Van

Regenmortel (2004) asserts that it becomes clearer that the specificity of a complex biological activity does not arise from the specificity of the individual molecules that are involved. In fact, the important developments in coherence quantum electrodynamics (CQED) came with the contention that condensed matter and living matter cannot be reduced only to their molecular components, but they can be reduced to the molecules oscillating in tune with an electromagnetic field (Del Guidice 1993; and Preparata 1995). However, for the same reason, we may argue that the organism as a whole may not be reduced to its constituent coherence subdomains. Particularly if we assume, following the CQED model, that the organism as a whole is an outcome of the collective dynamics of its constituent coherence subdomains at a certain matter information density and appropriate thermodynamic conditions.

In their article "No Entailing Laws, but Enablement in the Evolution of the Biosphere", Giuseppe Longo, Maël Montévil, and Stuart Kauffman (2012) stated that the aim of their article is to demonstrate that the mode of understanding in physics since Newton, namely differential equations, initial and boundary conditions, then integration that constitutes deduction, which in turn constitutes "entailment," fails fundamentally for the evolution of life. They argue that no law in the physical sense entails the evolution of life.

These points are at the heart of their considerations:

1)   In physics, the configuration space or phase space can be prestated. Dynamics are geodetics within

such prestated phase spaces, which may be very abstract, like in quantum mechanics.

2) In biological evolution, the phase space itself changes persistently. More so, it does so in ways that cannot be prestated.

3) Because we cannot prestate the ever-changing phase space of biological evolution, we have no settled relations by which we can write down the "equations of motion" of the ever-new biologically "relevant observables and parameters" revealed after the fact by selection acting on Kantian wholes in biological evolution, but that we cannot prestate. More, we cannot prestate the adaptive "niche" as a boundary condition, so we could not integrate the equations of motion even if we were to have them.

4) If the above is true, no law entails the evolution of the biosphere.

This is why, from the beginning, the founders of quantum mechanics were skeptical about the adequacy of quantum mechanics to explain life phenomenon. Bohr (1933) had the view that the distinction between living and nonliving systems was fundamental and a manifestation of his principle of complementarity. However, it was not clear at that time what characterizes such living and nonliving systems' complementarity. Answering the question regarding whether life is based on the laws of physics, Schrödinger says, "What I wish to make clear in this last chapter is, in short, that from all we have learnt about the structure of living matter, we must be prepared to find it working in a manner that cannot be reduced to the ordinary laws of physics. And that not on the ground that there is any 'new force' or what not, directing the behavior of the single atoms within a living organism, but because the construction is different from anything we have yet tested in the physical laboratory." He further emphasized the following: "We must be prepared to find a new type of physical law prevailing in it." And to the question concerning whether the new law could be nonphysical or super physical, he answered, "No. I do not think that. For the new principle that is involved is a genuinely physical one: it is, in my opinion, nothing else than the principle of quantum theory over again" Schrödinger (1944).

In 1949, Delbruck, inspired by Bohr, expressed a view that is similar to that of Schrödinger's: "Just as we find features of the atom, its stability, for instance, which are not reducible to mechanics, we may find features of the living cell which are not reducible to atomic physics, but whose appearance stands in a complementary relationship to those of atomic physics."

2.4—INTELLIGENT DESIGN

In the ontological domain, Ayala noted that reductionism versus antireductionism in its extreme form resolves into mechanism versus vitalism. The mechanists position is that organisms are ultimately made up of the same atoms that make up inanimate matter, and nothing more. Vitalists argue that organisms are made up of not only of material components (atoms and molecules and an aggregation of them) but also of some nonmaterial entity, variously called entelechy, vital force, vital energy, and the like.

At present, vitalsim has no distinguished proponents among biologists because it does not meet the requirements of a scientific hypothesis. However, recently a new form of vitalism was developed, called intelligent design movement, capitalizing mainly on the above-mentioned difficulties facing neo-Darwinism. Intelligent design is the assertion that certain features of the universe and of living things are best explained by an intelligent cause, not an undirected process such as natural selection. Consequently, intelligent design movement calls for an intelligent designer in order to account for the enormous growth of information and complexity characteristic of living systems. Intelligent design movement—Meyr (2005); Dembski (1998)—challenges naturalists to resolve the following:

- To propose a solution in which inanimate matter alone (unguided) develops information

- To determine where information comes from in the first place as well as how it can increase over time

Dembski tried to demonstrate in a mathematical way that what he calls "complex specified information" (CSI) cannot arise by natural causes. He says, "Natural causes are in principle incapable of explaining the origin of CSI. To be sure, natural causes can explain the flow of CSI, being ideally suited for transmitting already existing information. What natural causes could not do, however, is originate CSI. This strong proscriptive claim, that natural causes can only transmit CSI but never originate it, I call the law of conservation of information. It is this law that gives definite scientific content to the claim that CSI is intelligently caused . . . . Intelligent design entails that naturalism in all forms be rejected. Metaphysical naturalism, the view that undirected natural causes wholly govern the world, is to be rejected because it is false. Methodological naturalism, the view that for the sake of science, scientific explanation ought never exceed undirected natural causes" (Dembski 1998).

This author agrees with intelligent design movement concerning two issues. First, it is necessary to distinguish between physical information and bioinformation, for biocomplexity is not mere improbability. Biocomplexity is developmental functional complexity, or what Dembski calls CSI. However, numerous authors made this distinction (Adami 2002) without calling for a movement.

Second, I also agree with the claim that biological evolution is not a random process (i.e., it is goal directed). Having said this, I will try to demonstrate that biotic evolution and development are subject to a certain biological organizing principle (life principle) of physical and naturalistic origin, which accounts for the enormous growth of bioinformation ontogenetically as well as phylogenetically. The main issue is that intelligent design movement does not provide operational definition for bioinformation or (CSI) in order to find means to prove or substantiate its claim that the bioinformation cannot increase by naturalistic methods. Now it is possible to quantify bioinformation as fusion of energy and information and show that the origination and maximization of bioinformation is subject to a maximum action principle, according to which the increase of bioinformation is proportional to the organism's rate of change of action. Thus the nature of intelligent design is nature's intelligence.

CHAPTER 3

PARADIGM SHIFT

Despite the difficulties and limitations confronting self-organization models and quantum mechanics, these difficulties and limitations do not abandon or invalidate the need for organizing principles. In fact, Davies (2004) asserts that life being an emergent phenomenon exhibiting novel properties and principles is not in conflict with causal closure at the microscopic level. He argues that advances in cosmological theory suggesting an upper bound on the information processing capacity of the Universe (Lloyd 2002) may resolve this conflict for systems exceeding a certain threshold of complexity. A numerical estimate of the threshold places it at the level of a small protein. He indicates that this result may be traced in part to the operation of as-yet-to-be-elucidated biological organizing principles, consistent with, but not reducible to, the laws of physics operating at the microlevel.

To discover such a life-organizing principle would be our basic objective; however, overcoming the above-mentioned limitations and inadequacies, and revealing the hidden assumptions that obscure the discovery of life-organizing principle, necessitates a paradigm shift:

3.1—BROADENING THE CONCEPT OF INFORMATION

Information theory in its present formulation is inadequate to account for biosystems, despite its extensive usage in measuring correlations within DNA sequences. That is the case, as Adami (2002) put it, measuring correlations within a sequence is not going to reveal how the sequence is correlated to the environment within which it is to be interpreted. Moreover, bioinformation is not mere complexity in order to be measured only in bytes. The bioinformation is developmental functional complexity; in fact, bioinformation is about the phenotype (i.e., about expressing the genotype within a specific environmental context). And an adequate representation of bioinformation must be based on a general conceptualization of the nature of organism. In this regard, the organism is both an energy processor and an information system (Brooks 2001).

It follows that the first step toward the new theory is to make a distinction between physical information and bioinformation, or equally well between physical complexity and biocomplexity. Physical complexity has different measures; however, we limit our considerations to two of them: energy rate density and information, which are the most general and important ones. First: Energy rate density (erg/s/g) (Chaisson 2005) emphasized that starting with life's precursor molecules and proceeding all the way up to plants, animals, and brains, the same general trend typifies life forms such as for inanimate galaxies, stars, or planets; hence, the greater the complexity of a system, the greater the flow of energy density through that system. Second: Information (bits), complexity is measurable in terms of information units because a complex system is highly improbable. Again, there are different information approaches: Shannon, Kolmogorov, Schrödinger, and so forth. However, as Collier (2003) indicated, the different approaches complement each other rather than being competitors.

Now, neither of the two measures is eligible by itself to account for biocomplexity. The first measure (energy rate density) disregards biological systems as information systems. For the second measure, bioinformation is not mere complexity or improbability; bioinformation is developmental functional complexity. Consequently, a more general account of biocomplexity or bioinformation must take into consideration the fact that the biosystem is both an information system and an energy processor (Scaruffi 2003). To combine both aspects of bioinformation, the information stored in DNA and proteins is defined as developmental functional complexity within specific environmental context. This means the bioinformation is about the phenotype (i.e., about expressing the genotype within a given environmental context). It remains to be seen in the next chapter how the bioinformation can be represented in its phenotypic expression, as developmental functional complexity, having the dimensions of energy and information.

3.2—IDENTIFYING LIFE FRACTAL NATURE

A fractal is a rough or fragmented geometric shape that can be subdivided in parts, each of which, at least approximately, is a reduced size copy of the whole. As Mandelbrot (1982) indicates, a fractal is generally self-similar and independent of scale. Fractals model complex physical processes and dynamical systems. The underlying principle is that a simple process that goes through infinitely much iteration becomes a very complex process. Kurakin (2011) ascertains the ubiquitous nature of fractality in biological hierarchy, whereby certain organizational structures and processes are scale invariant and occur repeatedly on all scales of the biological hierarchy, at the molecular level, cellular, organism, population, and higher-order levels of biological organization. Based on his self-organization fractal theory (SOFT), Kurakin proposes the existence of universal principles governing self-organizational dynamics in a scale invariant manner. In this regard, two important features characterize the biological hierarchy as fractal:

First, numerous quarter-power scaling rules appear to span all levels of biological hierarchy, from molecules to ecosystems Niklas (2006). The most famous of these rules is Kleiber's (1947), which states that the basal metabolic rates scale as the three-quarter power of body mass. The second feature is the wide spread of Fibonacci sequence (1, 1, 2, 3, 5, 8, 13, 21, 34, 55 . . .) in biotic patterns (Knott 2001). When dividing two successive numbers of Fibonacci sequence, each by the number before it, especially at higher numbers, the result approaches 1.618, which is known as the golden ratio. A survey of zoological literature confirmed the wide occurrence of Fibonacci numbers in the organization of acellular and prokaryotic life forms as well as some eukaryotic protistans and in the embryonic development and adult forms of many living and fossil remains of metazoan animals Wille (2012). Wille also deduced from the correspondence between Hox gene number and animal segmentation that the Hox genes are themselves subject to the rule of Fibonacci numbers.

Jean-Claude Perez (1990) discovered the mathematical law that controls the self-organization of the basis T, C, A, and G inside DNA. He found that the consecutive sets of DNA are organized in sets of distant order, called resonance. The resonance represents the special proportion of DNA parts pursuant to Fibonacci numbers. For example, if 144 adjacent nucleotides of DNA formed from 55 bases of T and 89 bases of A, C, or G, then the proportion (55-89-144) represents the resonance. Moreover, each full cycle of the DNA helix spiral measures 34 angstroms in length and is 21 angstroms wide, whereby 34 and 21 are Fibonacci numbers. It is remarkable that the DNA fulfills both requirements of scale invariance and golden ratio based on fractality by having nested dodeca-icosa structure. The dodeca consists of 12 pentagonal faces and the icosahedron of 20 triangles. These shapes can all mathematically turn into one another, and this transformation takes place with ratios linked to the golden ratio. The dodeca and its dual icosa nest infinitely in 3 dimensions, where each vertex x, y, z coordinate is a simple whole exponent of golden ratio, meaning that the distance to the center in this nodal array from every node is a simple multiple of golden ratio, Winter(2012).

It is clear from this discussion that Fibonacci numbers, or the golden ratio, is the basis of the natural fractal design of living systems. Therefore, it is tempting to ask whether this design rests on sound physical laws or if it is just a coincidence. In fact, there are some physical reasons; we would argue that the appearance of golden ratio in biotic patterns is a deterministic consequence of the maximum action principle, which is the driving force of biotic evolution and development. Moreover, Dan Winter (2012)—a pioneer on golden ratio in physics—asserts that golden ratio fractality is a condition of recursive constructive interference. In his view, golden ratio fractality is the only geometry that allows wave patterns to add and multiply recursively and constructively, thus producing optimum charge distribution and coherence. He coined the term quantum fractal field to designate the state of perfected charge distribution and coherence characteristic of the DNA. I would like to add and show that the DNA or genome is a self-replicating quantum information fractal field that generates, in addition to weak electromagnetic vibrations, bioinformation oscillations. Moreover, I demonstrate that the functional stationary quantum states of living systems, ontogenetically as well as phylogenetically, correspond to the optimum coherence states.

3.3—EXTENSION OF QUANTUM FIELD THEORY

At the end of the nineteenth century, light was thought to consist of waves of electromagnetic fields that propagated according to Maxwell's equations, while matter was thought to consist of localized particles. This division was challenged when, in his 1905 paper on the photoelectric effect, Albert Einstein postulated that light was emitted and absorbed as localized packets, or "quanta" (now called photons). These quanta would have an energy

$$E = hv \quad (1)$$

where v is the frequency of the light and h is Planck's constant. Robert Millikan and Arthur Compton confirmed Einstein's postulate experimentally over the next two decades. Thus it became apparent that light has both wavelike and particle-like properties. On the other hand, the concept of matter waves, or de Broglie waves, reflects the wave-particle duality of matter. Louis de Broglie proposed the theory in his 1924 PhD thesis. The de Broglie relations show that the wavelength is inversely proportional to the momentum of a particle and is also called de Broglie wavelength. In addition, the frequency of matter waves, as deduced by de Broglie, is directly proportional to the particle's total energy (i.e., the sum of the particle's kinetic energy and rest energy). In 1926, Erwin Schrödinger published an equation describing how this matter wave should evolve and used it to derive the energy spectrum of hydrogen.

Based upon this physical realization of the universe, it is apparent that an important property of matter has been missed, which renders contemporary physical theory impotent to account for life phenomenon. Matter has not only matter waves microscopally, at high-mass density, but it also has bioinformation oscillations at high-information density. Consequently, quantum theory is not applicable to biosystems in its present formulation, because its phase space does not contain fundamental biological variables. This is why the specificity of a complex biological activity does not arise from the specificity of the individual molecules that are involved (Van Regenmortel 2004). Moreover, the interactions of a classical system with its environment decohere the wave function (Davies 2004). To overcome these limitations, something new is needed. It is necessary to broaden the concept of quantum field. We propose that the DNA or genome represents a new type of quantum information fractal field (QIFF), which has the following properties:

-   It is a product of the collective dynamics of ordinary physical fields optimized by golden ratio based fractal geometry.

- It generates bioinformation oscillations through successive generations (overcoming reductionism barrier by broadening the ontological foundation of contemporary physical theory).

- It is subject to maximum coherence and optimum distribution of charge, generated by golden ratio based fractal geometry (overcoming decoherence barrier).

- It is a function over bioinformation and time, so it is relatively independent of phase space coordinates (overcoming phase space coordinates barrier).

- It is subject to a maximum action principle in order to account for self-organization, self-replication, and self-evolution (overcoming thermodynamic barrier).

The biological hierarchy being fractal (i.e., displaying the property of self-similarity and complexity and incorporating the golden ratio) has a lower level consisting of elementary particles, atoms, and molecules as well as an upper level consisting of biomolecules, cells, organisms, and ecosystems. The upper level has the emergent property of QIFF, which is irreducible to the lower level. Hence self-similarity and irreducible complexity characterizes the biological hierarchy.

Now, in a similar manner, we would also like to assume that the hierarchy of physical laws that describe the biological hierarchy is fractal. Consequently, to account for the biological hierarchy, the hierarchy of physical laws must also be self-similar and irreducibly complex. We know that the lower level of the hierarchy of physical laws is subject to quantum field theory for which the wave function (i.e., Schrödinger system) is the basic attribute. So how can we reconcile the quantum self-similarity of the hierarchy of physical laws with its irreducible complexity? This dichotomy could only be resolved by a new generalized irreducible quantum field theory, which we call QIFFT. However, since the Schrödinger system is the basic characteristic of quantum field theory, it is reasonable to argue that the new quantum information fractal field theory (QIFFT) would also be characterized by a generalized irreducible Schrödinger type of system. We call such a generalized Schrödinger type of system the life-organizing principle or just life principle.

But what do we mean by a generalized Schrödinger type of system, or the life-organizing principle? In general, and this is what is to be revealed, the life-organizing principle is an attribute of a quantum information fractal field (QIFF) that is a function over bioinformation and time. The quantum information fractal field is characterized by bioinformation that has the dimensions of energy and information, meaning active information, whereas ordinary quantum field is characterized by passive information. In other words, while an ordinary quantum field carries information, the QIFF generates bioinformation. Moreover, the life-organizing principle must satisfy the following conditions:

- Since the eigen value of the Hamiltonian operator acting on a Schrödinger system of an inanimate system represents the system's total energy, while the eigen value of the proposed generalized Hamiltonian analog acting on the life principle represents not only the system's total energy but also its genome physical information and lifetime (i.e., its bioinformation or vitality). It follows total energy is sufficient to contain the dynamical essence of an inanimate system, while vitality which is a fusion of matter-energy, information and time is more appropriate to characterize the dynamical essence of a living system.

- The life principle must admit limiting transition to linear reversible quantum mechanics.

3.4—EXTENSION OF ACTION PRINCIPLE

Action is an attribute of the dynamics of a physical system; it is a mathematical functional that maps the system's path and its argument to a real number. The action can be represented by an integral over time, taken along the path of the system, specifying the initial and final time evolution:

$$S = \int L dt \quad (2)$$

S is the action and t time. The Lagrangian (L) is given by:

$$L = T - V \quad (3)$$

T and V, respectively, are the kinetic and potential energy of the system.

The action has the dimension of energy X time; its unit is Joule second, in the International System of Units. Classical mechanics postulates that the path followed by a physical system is that of minimum action or more strictly is stationary.

The classical equations of motion of a system can be derived from the principle of least action. The principle led to the development of Lagrangian and Hamiltonian formulations of classical mechanics. The principle remains central in modern physics and mathematics. It is applied to the theory of relativity, quantum mechanics, and quantum field theory. It also becomes a focus of modern mathematical investigation in Morse theory. Thus the action principle is the most powerful tool of physics, from which almost all the fundamental laws of physics can be derived.

It is tempting to ask about the impact of such a fundamental principle on living systems. It is apparent that the spontaneous growth, development, and active function of living systems cannot be an expression of least action. However, since the mathematical form of the action principle allows endpoints corresponding to the maximum of action, present-day physics does not utilize the full potential of the action principle. Grandpierre (2007) argues that this situation comes because of the nature of the physical problems to which the endpoints of the action principle is fixed by the initial conditions. Moreover, physical behavior in most cases corresponds to the minimal form of the action principle. Grandpierre, then, proposed that the first principle of biology arises at the other extremum of integrated action that corresponds to the maximum of integrated action. Furthermore, Grandpierre proposed a mathematical form for a path integral of most action. His approach resides in applying the Lagrangian using extropy, which measures the entropic distance from thermodynamic equilibrium.

We share with Grandpierre the conviction that the enigmatic spontaneous growth, development, and functional activity of living systems is most probably an expression of a fundamental principle of maximum action. In other words, the dynamical essence of living systems may be imbedded within a path of maximum action. Indeed, a general requirement for a biological system to maintain a path of maximum action is that its rate of change of action must increase, at least initially, in order to match the path or trajectory of maximum action. To achieve this goal, we employ the phase of the genome's bioinformation oscillations as represented by the life-organizing principle. It is interesting to find out that the phase in the life-organizing principle (equation 18) represents a biological path of maximum action.

CHAPTER 4

QUANTUM INFORMATION FRACTAL FIELD THEORY (QIFFT)

4.1—POSTULATES OF QIFFT

The concept of life-organizing principle (or life function) was introduced by Elsheikh (2010, 1999, 1988), and it is, in essence, an expression of the equation of motion that reveals genome dynamics for successive generations. To uncover such an equation of motion, we need to have fundamental biological variables that which contain the dynamical essence of the system. For this sake, we propose here the following QIFFT postulates:

I. A living system's genome is a self-organizing, self-replicating, and self-evolving quantum information fractal field, QIFF.

II. The QIFF generates, in addition to weak EM waves, bioinformation oscillations through successive generations.

III. The bioinformation oscillations contain the dynamical essence of the living system.

IV. The bioinformation sustains the living state.

There are sufficient reasons to claim that the DNA or genome is a quantum information fractal field:

- Quantum aspect: Life is a fundamentally discrete phenomenon (i.e., discrete nucleotides, discrete genes, discrete chromosomes, discrete cells, heredity, evolution, and so forth).

- Information aspect: For life to exist, an information system is needed to produce and regulate life functions. Thus the DNA or genome is an information-storing, processing, and replicating system.

- Fractal aspect: Jean-Claude Perez (2010) and Dan Winter (2012) demonstrated that the genome has golden mean ratio based fractal structure.

- Field aspect: Whether classical or quantum, a field is defined as a function over space and time. This definition is not sufficient to account for the genome as a field, because a living system's dynamics or functionality depends on its bioinformation or biocomplexity rather than on the space coordinates it occupies. So we define the genome field as a function over bioinformation and time. Such field generates, in addition to weak EM waves, bioinformation oscillations. As stated earlier, the genome field is nothing other than the outcome of the collective dynamics of ordinary physical fields optimized by golden ratio based DNA fractal geometry. Consequently, the genome field is subject to a maximum action principle that characterizes its self-organization, self-replication, and evolution. Note that the maximum action principle is not a property of an ordinary physical field. This is why life has always been problematic from the perspective of ordinary physics. Regarding DNA EM wave emission, Montagnier (2010) discovered such waves, while Elsheikh and Winter (2013), in an article entitled "Theoretical Basis for Montagnier's Revolutionary Discovery," which is also included in this book, propose fractal field phase conjugation as mechanism for the wave emission.

4.2—BIOINFORMATION (VITALITY)

We claim that the above-mentioned postulates are appropriate to define life and provide basis for its mathematical representation. To establish a quantitative measure of bioinformation and to demonstrate that the genome generates bioinformation oscillations in accordance with these postulates and vision, we first consider some findings of information biologists.

Information biologists have been able to give a hierarchical account of information in living systems (Galtin 1972; Brooks and Wiley 1988). Brooks and Wiley found that even heuristic simulations emulating biological processes associated with storage and transmission of information (i.e., reproduction, ontogeny, and speciation) produce three important generalities:

i. H_{obs} is an increasing function of time.

ii. H_{obs} is a concave function of time as historical constraints retard the rate of entropy increases.

iii. The difference between H_{max} and H_{obs} is an increasing function of time, permitting the growth of structure and organization on very long time scales. However, on short time scales, Smith (1999) demonstrated that $(H_{max} - H_{obs})$ is a difference between two increasing convex functions that grows monotonically up to adulthood and then stays stable (homeostasis) or decreases (senescence).

Therein:

H_{obs} = information content, or constrained complexity. It is the actual entropy calculated on the basis of the observed distribution of components of the system.

H_{max} = information capacity, or complexity, estimated by calculating the entropy value for the components of the system at any time if they were all randomized.

$H_{max} - H_{obs}$ = macroscopic information, or information.

Since $(H_{max} - H_{obs})$, the information assigned to a growing organism, according to Smith, is an increasing function of time for $0 \le t \le \alpha$, where α is the time when the organism is fully grown (i.e., adult). Afterward biological information, $(H_{max} - H_{obs})$, decreases for $t > \alpha$. Let us in a general case define the information assigned to a growing organism by $I(t)$ and then assume the following:

$$I(t) \propto (H_{max} - H_{obs}) \quad (4)$$

Therefore $I(t)$, has the following properties:

$$\dot{I}(0) > 0$$

$$\dot{I}(\alpha) = 0 \quad (5)$$

$$\ddot{I}(\alpha) < 0$$

Thus the organism has maximum information at adulthood.

Inspired by these experimental findings of Brooks, Wiley, and Smith, we construct a model for bioinformation, starting with introducing the following definitions:

Definition 1:

Bioinformation, I(t), which is the information stored in DNA and proteins, is developmental functional complexity, within a specific environmental context, where (t) is the time measured from the moment of initial growth.

Definition 2:

Vitality, v (t), is defined as the genome capacity to generate developmental functional complexity (the capacity to generate bioinformation or phenotype).

Thus capitalizing on the fourth postulate, it is possible to recognize the fundamental biological variables. The postulate emphasizes that bioinformation sustains the living state, by which we mean:

a) Maintaining the system's survival

b) Maintaining the systems self-propagation

By maintaining the system's survival, we mean using a certain matter-energy growth function to exhibit the basic properties of life during a certain period of time. We usually call such a period a life span, A, or life expectancy. Accordingly, we restrict the fundamental bioinformation variables to the following:

X—Matter-energy growth function, E(t), (i.e., total matter-energy metabolized by the system, measured in calories)

Y—Life expectancy, ℓ. (ℓ = A − t), where A is the life span, measured in minutes, hours, days, or years in accordance with the organism.

Z—Self-propagation, or natality rate, N(t)

The effects of all other aspects of biological information, in addition to the effects of environmental conditions, are to be interpreted as a perturbation or a modification of these variables, [E, ℓ, N]. Note that [E, ℓ, N] are average values for a given organism of a given species.

Having specified and quantified the bioinformation fundamental variables, we proceed to attempt to quantify vitality in terms of these variables:

$\therefore$ I(t) = I [E(t), ℓ(t), N(t)]   (6)

If there is no explicit natality, N(t), dependence, we get:

I(t) = I [E(t), ℓ(t)]

Consider the simple plausible assumption:

i – I(t) $\propto$ E(t), ii – I(t) $\propto$ ℓ^a

It follows from i and ii,

I(t) = bE(t) ℓ^a (7)

Therein, b is proportionality constant which has information units. It might be a good idea to suggest that the proportionality constant, b, represents the given species genome's physical complexity measured in bits or bytes. And a is a parameter which depends on the species and it is assumed to relax the dependence on ℓ. For the nonexplicit dependence on N(t), we let I(t) = v(t), and then we get:

$\therefore$ v (t) = bE(t) ℓ^a = bE(t)(A - t)^a (8)

We call v (t), vitality. It follows, then, that the phenotypic expression of bioinformation characterizes the physical nature of vitality. Being a function of the phenotypic fundamental variables, vitality measures the system capacity to generate bioinformation. For simplicity, we limit our considerations to energy metabolism, where:

$$E(t) = \int_0^t M(x)dx, \quad (9)$$

M(t) is the metabolic rate, usually given by:

M(t) = $cW^r(t)$, c and r (= ¾) are constants, and W(t) is the organism's body mass, (Kleiber 1947); b is a proportionality constant assumed to represent the organism's nonredundant functional genome size, measured in bytes.

Result 1

In our present considerations, vitality being a product of an increasing function of time E(t) and a decreasing function of time, $\ell^a = (A-t)^a$, it has the following properties:

i) $\dot{v}(0) > 0$

ii) $\dot{v}(\alpha) = 0$ (10)

iii) $\ddot{v}(\alpha) < 0$

iv)   v(A) = 0

Consequently, in this model, there exists a vitality function, v(t), that satisfies the following condition:

It increases before adulthood, reaches a maximum at adulthood, decreases afterward, and becomes zero when the organism dies. (See appendix for proof of (10)).

Using dimensionless variables (see appendix), we can draw the vitality curve:

Table 1

x	0	1	2	3	4	5	6	7	8
y	4096	6482.7	9590.4	12562.5	13759.2	8845.2	2620.8	218.4	0

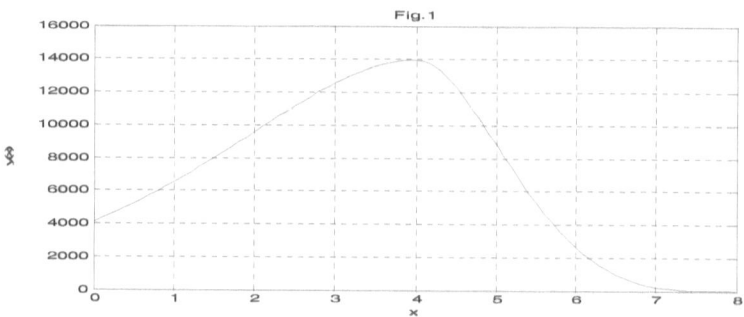

Fig. 1. Dimensionless representation of the vitality function based on the special case of exponential growth functions. The x-axis represents time, and the y-axis represents vitality.

Note: it is evident, no matter what energy growth function the organism has, that the present model always determines a vitality curve for it.

Now, to compare or measure the biocomplexity or bioinformation of different species, using equation (8), we get:

$$v_s(\alpha) = A^{-a}v(\alpha) = bA^{-a}E(\alpha)\ell^a(\alpha) \quad (11)$$

Therein, $v_s(\alpha)$ is life span specific vitality, the unit of which is bytes x calories. We compare the biocomplexity of different species by considering their biocomplexities at adulthood. Consequently, as evident from (11), the organism's bioinformation or biocomplexity has information as well as energy dimensions. In this manner, biocomplexity, being developmental functional complexity, is distinguished from mere improbability. Ulanowicz (2004) revealed that biological organization has the dimensions of energy and information and is measurable in terms of ascendency units (flow x bits).

4.3—BIOINFORMATION OSCILLATIONS

The model may be usefully employed to discuss vitality for successive generations. We shall essentially be concerned with unicellular organisms, particularly those that reproduce by binary fission. For such systems, figure 1 represents the average vitality for one generation (i.e., when the organism grows and then dies). However, a unicellular organism usually does not die; it starts to divide when it is fully grown. Referring to figure 1, division occurs at time $t = A_1$, where $\alpha < A_1 < A$. The parent cell gives birth to two identical daughter cells so that $v(A_1) = 2v(0)$, where $v(0)$ is the initial vitality. Again each daughter cell grows and divides in the same manner. Thus the average vitality function, vitality per unit cell or organism, for successive generations, v_g, under constant environmental conditions, is a periodic function of time with period A (assuming no confusion, we have deleted the subscript from A_1).

Then this follows:

$$v_g = v(t + mA) = v(t) \qquad (12)$$

Therein, m = 1, 2, 3 . . . and is the number of cell divisions or generations. Equation (12) defines vitality as a periodic function of time—that is, it represents vitality oscillations or, equally acceptable, the bioinformation oscillations generated by the genome as a self-replicating quantum information fractal field.

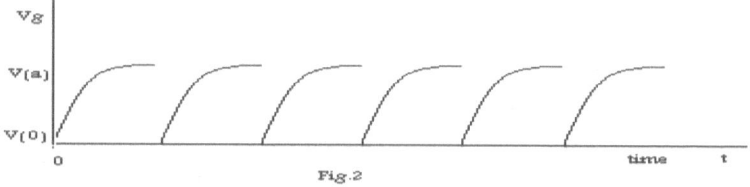

Fig.2

Fig. 2. Hypothetical representation of bioinformation oscillations of a unicellular organism for successive generations. The x-axis represents time, and the y-axis represents vitality.

Figure 1 describes the situation if the unicellular organism completes the cell cycle and dies instead of dividing (i.e., the figure determines its vitality curve). However, for figure 2 to take place, the organism must divide successively—that is, must have a natality rate N(t). Accordingly, we shall regard figure 2 as representative of the life state. It is also important to note that v(t) is a sectionally continuous function. Such a function can be represented by a Fourier series, which is convergent where v(t) is convergent and converges to the arithmetic mean of the values approached by v(t) from the right and the left of the discontinuity. Since this arithmetic mean, in the present model, is 3v(0)/2, it is desirable to define a new function, s(t), as follows :

$$s(t) \qquad = v(t) \quad , \qquad 0 \le t < A$$

$$\tag{13}$$

$$= 3v(0)/2 \quad , \qquad t = A$$

Having stated this, assuming no confusion, we discuss v(t) as a continuous function, whereas we actually mean s(t).

Consequently, the genome's bioinformation measured in calories x bits oscillates in the time-vitality (t-v) state space during successive generations. This model is equally applicable to multicellular organisms which reveal an overlap of generations (i.e., overlap of bioinformation cycles or oscillations).

The following definitions will be useful in our subsequent considerations.

Instantaneous total vitality $V(t) = \int_0^t v(x)\, dx$ (14)

Total vitality $V(A) = \int_0^A v(t)\, dt$ (15)

Genome's total bioinformation, T(A), is given by:

T(A) =

$$\int_0^A I[v(t), N(t)]dt = \int_0^A v(t)dt + \int_0^A f(N(t))dt = V(A) + F(N(A)) \quad (16)$$

N(t) = (dP/dt)/P is natality rate; P is population size; f(N(t)) is natality density function; V(A) is total vitality; F(N(A)) is total natality density function.

4.4—THE LIFE-ORGANIZING PRINCIPLE

Based on the periodicity of v(t) and the third postulate of QIFFT, we assume that the life state of an organism can be described by a complex function, L(v, t), corresponding to the bioinformation oscillations, v(t), which we call life-organizing principle. Also based on the complementarity of matter waves and bioinformation oscillations mentioned earlier, a material system, animate or inanimate, does not simultaneously possess both matter waves and bioinformation oscillations descriptions. It is appropriate to seek the life-organizing principle formulation in terms of the QIFF, $L(v, t)$, independent of ordinary quantum field representation. Hence with regard to the two complementary properties of matter, matter waves and bioinformation oscillations, let us consider the phase of Schrödinger's system-wave function—a black box. It follows that either the black box could be loaded by matter waves to describe quantum mechanical phenomena or it could be loaded by bioinformation oscillations in order to describe biological phenomena. In the latter case, it is called life-organizing principle. Note that the life-organizing principle does not contain Planck's constant; rather, it contains an analog of it (i.e., a new constant having the same dimensionality as Planck's constant). In addition, based on the above-mentioned fractal nature of the hierarchy of physical laws, which necessitates some sort of generalized irreducible Schrödinger type of system, the life-organizing principle should satisfy the following conditions:

I. $L = L(v, t)$.

II. L is a Schrödinger's type of system.

III. L is a nonconservative, nonlinear, and irreversible system, conditions that account for the structural stability characteristic of biological systems.

We shall therefore assume the simple following form:

$$L = L_o e^{i\Phi/G} \quad (17)$$

L_o is the amplitude; G is a constant to be found and has the dimension of Planck's constant. To satisfy the above-mentioned conditions, we limit our considerations to the concrete example in which Φ is given by:

$$\Phi(t) = \int_0^t E(1 - \frac{x}{A})^a \, dx \quad (18)$$

This completes the definition of L.

Note: from (8), the instantaneous total vitality can be given by:

$$V(t) = \int_0^t v(x) \, dx = b\int_0^t E(x)(A-x)^a dx \quad (19)$$

It follows that using equations (8), (14), (18), and (19), we obtain:

$$\Phi(t) = b^{-1}A^{-a} V(t) = b^{-1} A^{-a} \int_0^t v(x) \, dx$$

$$\therefore \quad L = L(0) \exp. \frac{i}{bA^a G} \int_0^t v(x) \, dx$$

$$\therefore L = L(0) \exp. \frac{i}{k} \int_0^t v(x) \, dx \quad (20)$$

Therein, $k = bA^a G$ (21)

$$\therefore \quad \ddot{L} - \frac{\dot{v}}{v} \dot{L} + \frac{v^2}{k^2} L = 0 \quad (22)$$

Equation (22) is an analog of Schrödinger equation, but it is different (other than being a system of bioinformation) in that it is a generalized Lienard's system, which is a nonlinear, nonconservative, and irreversible system that admits limit cycle. Irreversibility, being a basic difference between life phenomenon and microphysical quantum laws, which are reversible, it is clear that the vitality function equation (8) does not accept any change of (t) for (−t). Such a change deprives vitality from its biological properties and ultimately explodes the function. Stable limit cycle solutions usually characterize structural stability, a property of high significance to biosystems. It is also appropriate to remark that in addition to representing the phase of genome's bioinformation oscillations, $\Phi(t)$ defines a biological path of maximum action (i.e., the life-organizing principle maximizes action, as we would like to demonstrate).

Definition 4.1:

The generalized Hamiltonian operator analog $\hat{D}$ is defined by:

$$\hat{D} = -ik\frac{d}{dt} \quad (23)$$

Result 2:

Vitality is the eigen value of the generalized Hamiltonian.

Proof:

From equation (20), we get:

$$\hat{D}L = v(t)L \quad (24)$$

As we said earlier, vitality, being a product of the system's total matter-energy metabolized, genome physical information and life expectancy, is more appropriate to contain the dynamical essence of a living system.

Definition 4.2:

We call matter-waves versus bioinformation oscillations, generalized complementarity.

Note: Bohr's complementarity is particle-wave complementarity, whereas generalized complementarity is living nonliving complementarity. The living system is characterized by bioinformation oscillations, while the nonliving system is characterized by matter waves. Also note that macroscopic nonliving systems (having no bioinformation oscillations) for which matter waves are negligible are reducible to the ordinary laws of physics; this is why classical mechanics is a special case of quantum mechanics.

Definition 4.3:

We call the extended domain of physical theory, which is a consequence of generalized complementarity, generalized physics, or QIFFT. What we have in mind is that the elimination of irreversibility from the proposed quantum-information fractal biology may yield the basic laws of linear reversible quantum mechanics in the limiting case.

Based on the generalized complementarity, no material system possesses simultaneously both matter waves and bioinformation oscillations for its description. It follows that life is irreducible to ordinary physics, while it is reducible to QIFFT.

Self-Sustained Oscillations

Nonlinear nonconservative systems that admit limit cycle solution are said to have self-sustained oscillations and are consequently structurally stable. They preserve the topology of their trajectory under sufficiently small perturbations. The behavior of a system undergoing sustained oscillations is stationary in the sense that it is repeating the same pattern over and over again. Even though the system itself is not stable in the state space, its behavior pattern is stationary. A common feature of oscillations of this kind is that their stationary oscillatory state does not depend on the initial conditions, as for conservative systems, but depends on the parameters of the system, which means it is determined by the differential equation itself. Self-sustained oscillations cover wide phenomena that range from physics to chemistry to biology.

Minorsky (1962) emphasized that "limit cycles and in particular stable limit cycles, are fundamental in the theory of oscillations of nonlinear nonconservative systems—the only kinds of systems in which they can arise." But why are self-sustained oscillations characteristic of nonlinear nonconservative systems? Migulin (1983) answered this question by showing that a simple self-oscillating system could be represented by storage element and a feedback channel with constant source. If we denote the stored oscillatory energy in the system by E, then in steady mode of self-oscillations, the changes of the oscillatory energy during a period will, by definition, be equal to zero:

$$E_{T+t} - E_T = 0 \quad or \ (\Delta \overline{E})_T = 0$$

∴For a conservative system: $\dot{E} = 0$

While in dissipative systems: $\dot{E} = -F(t) < 0$ (25)

F(t) is the function characterizing the dissipative properties of the system. For dissipative systems F(t) > 0, the dissipation function defines the power losses in the system. Usually F(t) > 0, but in self-oscillating systems, time intervals with certain amplitudes and speeds are possible at which F(t) < 0. It is evident that the condition F(t) < 0 is inherent in only active systems. From here, Migulin stated the basic equation of energy balance for self-oscillating systems:

$$\int_0^T F(t)\,dt = 0 \quad (26)$$

Accordingly, in linear and nonlinear dissipative systems, a self-oscillatory process is impossible. Thus for self-sustained oscillations to exist, it is required that the dissipation function F(t) be alternating in sign. In this case, the oscillatory energy increases during one part of a period, which can be described with known phenomena of negative resistance or damping, while during the other part of the period, the oscillatory energy decreases. This ensures the energy balance, which means that steady state oscillations exist in the system.

The Life-Organizing Principle Admits Limit Cycle:

Since the life-organizing principle, equation (22), represents a nonlinear, nonconservative, and irreversible system, which describes self-sustained bioinformation oscillations, and it represents a generalized Lienard's type of system. It follows

93

the LOP admits limit cycle solution, and the living system is structurally stable:

$$\ddot{x} + F(v,\dot{v})\dot{x} + G(x,v) = 0 \quad (27)$$

Equation (27) is a generalized Lienard's system, which can be satisfied by (22) as follows:

i. Lienard's condition: $F(v,\dot{v}) < 0$, for $0 \le t < \alpha$;
$F(v(\alpha), \dot{v}(\alpha)) = 0$, for $t = \alpha$;
$F(v,\dot{v}) > 0$, for $\alpha < t \le A$.

ii. The damping coefficient, $F(v,\dot{v})$, is periodic and possesses explicitly the vitality dissipation function, $\dot{v}$, with alternating sign. Note: $\dot{v} = \dfrac{dv}{dt}$, $\dot{x} = \dfrac{dx}{dt}$.

Where x = x [v (t)], subject to the eigen value condition: $\dot{x} = ivx$ and v(t) = v(t+A), we thus get the next result:

Result 3:

The life-organizing principle (LOP) admits limit cycle if the integral of the damping coefficient over the whole cycle vanishes.

Proof: From equation (22) and (27), we get:

$$\int_0^A F(v,\dot{v})dt = -\int_0^A \frac{\dot{v}}{v}dt = \ln v(0) - \ln v(A) \quad (28)$$

From (10), v(A) = 0

Normalizing v(0) = 1, yields: $\displaystyle\int_0^A F(v,\dot{v})dt = 0 \quad (29)$

Therefore, (29) proves our result.

Using (20), the limit cycle is then given by:

$$\lim_{t \to \alpha} it \frac{-ik}{L} \frac{dL}{dt} = v(\alpha) \quad (30)$$

This result is intuitively based on the fact that the eigen value of the Hamiltonian ($i\hbar \frac{d}{dt}$) of a conservative system defines elliptical trajectory. Likewise, the eigen value of LOP, $v(\alpha)$, is supposed to define its limit cycle.

One of the important models of biological oscillators that later on found extensive study in literature is the work of Goodwin (1963). Goodwin suggested that cell division implies an autonomous oscillating signal, so he discussed several aspects of negative feedback control processes. In particular, he proposed certain models for protein synthesis that might exhibit oscillatory properties. Niclos and Prigogine (1977) showed that some of the most spectacular aspects of biological activity, such as control of cellular division, or cellular differentiation and morphogenesis, could be modeled as self-oscillating phenomenon.

A limit cycle is also called an attractor, a set of states of a dynamic physical system toward which that system tends to evolve, regardless of the initial conditions of the system. There are different types of attractors other than the limit cycle (e.g., a point attractor that is an attractor consisting of a single state, such as the final states of a falling pebble or damped pendulum). Another example is a strange attractor, which is an attractor for which the evolution through the set of possible physical states is no periodic, resulting in an evolution through a set of states defining a fractal set. Note: an attractor is only effective within a certain area of space, which is called its basin. In general basins of attraction are areas in the state space of a system where, if an object enters the basin, it will "slide down" the basin to the attractor at its bottom, and stay there.

Definition 4.4:

We call the life-organizing principle that describes muticellular organism dynamics a major attractor, while that which describes cellular dynamics is a minor attractor.

CHAPTER 5

MAXIMUM ACTION PRINCIPLE

5.1—PATH OF MAXIMUM ACTION

To account for the spontaneous growth, development, and functional activity of living systems, the living system must maintain a path of maximum action. Under such circumstances, the genome capacity to generate developmental functional complexity (vitality) must be correlated to the rate of change of action to match the path. Thus we try to demonstrate that the phase of the genome's bioinformation oscillations, which has action units, is the path of maximum action we are looking for. Therefore, let us start from equation (18):

$$\Phi(t) = \int_0^t E(x)(1-\frac{x}{A})^a dx \quad = A^{-a}\int_0^t E(x)(A-x)^a dx \quad (31)$$

From equation (8):

$$\Phi(t) = b^{-1}A^{-a}\int_0^t v(x)dx \quad = K\int_0^t v(x)dx \quad (32)$$

$$K = b^{-1}A^{-a} \quad (33)$$

We propose equation (32) to be the biological path of maximum action, which is a time-dependent integral. The integral being time dependent simplifies finding the trajectory or equation of motion, v(t), without resorting to variational methods. Note:

$$\Phi(A) = K\int_0^A v(t)dt = \text{organism total action} \quad (34)$$

Result 5.1:

$\Phi(A)$ is an integral of maximum action.

Proof:

Given equation (32):

$$\Phi(t) = K \int_0^t v(x)dx$$

$$\therefore \dot{\Phi}(t) = K v(t) \quad (35)$$

$$\therefore \dot{\Phi}(A) = Kv(A) = 0 \quad (36)$$

$$\therefore \Phi(A) = \text{constant}$$

It follows, from (36), that t = A is a critical point. To clarify whether the critical point is a maximum or minimum, we consider the second derivative:

$$\ddot{\Phi}(t) = K(\dot{E}(t)(A-t)^a - aE(t)(A-t)^{a-1}) \quad (37)$$

$$\ddot{\Phi}(A) = K(0) = 0 \quad (38)$$

From (38), the critical point is an inflection point at t = A (i.e., when the organism is dead). The inflection point indicates that there is a change of state from maximum action to minimum action, or vice versa. However, it is reasonable to assume that the organism changes from the living state of maximum action to the dead state of minimum action. Let us check this assertion by investigating the behavior of the trajectory just before death.

Let t = $\varepsilon \approx A \Rightarrow A - \varepsilon = \delta > 0$ (39)

Substituting (39) in (37) yields:

$$\therefore \ddot{\Phi}(\varepsilon) = K\left[\dot{E}(\varepsilon)(A-\varepsilon)^a - aE(\varepsilon)(A-\varepsilon)^{a-1}\right] \quad (40)$$

$$= K\left[\dot{E}(\varepsilon)\delta^a - aE(\varepsilon)\delta^{a-1}\right] \quad < 0$$

$K > 0$, $\dot{E}(\varepsilon) < 0$, $\quad a > 1$ $\quad$ (41)

Note: $\dot{E}(\varepsilon)$ is the metabolic rate at old age, which is negative.

Equation (40) proves this:

$$\ddot{\Phi}(\varepsilon) \approx \ddot{\Phi}(A) < 0 \quad (42)$$

This indicates that the trajectory is concave downward everywhere, which forces t = A to produce a maximum.

$$\therefore \Phi(A) = K \int_0^A v(t)dt = cons\tan t = \max imum \quad (43)$$

5.2—FIRST LAW OF SELF-ORGANIZATION

We indicated earlier that for a living system to grow and develop spontaneously and function actively, it has to maintain a path of maximum action. Consequently, the living system rate of change of action must increase so that the system can match the maximum action path.

Let us start from equation (35):

$$\therefore \dot{\Phi}(t) = K v(t)$$

Equation (35) shows that the rate of change of action increase is proportional to vitality, which is the genome capacity to generate developmental functional complexity. We call (35) the first law of self-organization. It is the driving force of self-organization—that is, it drives the zygote from the initial phase of relative thermodynamic equilibrium and relative maximum entropy to the adult phase of far from equilibrium and minimum entropy. It is appropriate to mention that Kurakin expressed the same law qualitatively when he wrote, "Third, the degree of complexity and order within a self-organizing nonequilibrium system and the rate of energy/ matter passing through the system correlate in a mutually defining manner" (Kurakin 2011). On the other hand, the kinetics of autocatalytic replication (Pross 2003) and fractal field phase conjugation (Winter 2012) provide the mechanism for such maximization of action.

For equation (35) to represent a path of maximum action, $\ddot{\Phi}(t)$ must satisfy the following conditions:

$$
\begin{aligned}
\ddot{\Phi}(0) & > 0 \\
\ddot{\Phi}(\alpha) & = 0 \\
\dddot{\Phi}(\alpha) & < 0 \\
\dot{\Phi}(A) & = 0
\end{aligned}
\qquad (44)
$$

Conditions (44) are based on the conditions imposed on vitality by equation (10), which have proof in the appendix. Equations (44) indicate that the organism rate of change of action increases before adulthood and has a maximum when the organism is fully grown. It decreases afterward and becomes zero when the organism dies.

Since equation (11) offers an operational definition for comparing the biocomplexity of different species, we employ it using equations (33) and (35) to redefine biocomplexity in terms of the rate of change of action:

$$
v_s(\alpha) = A^{-a} v(\alpha) = A^{-a} \frac{\dot{\Phi}(\alpha)}{K} = \dot{\Phi}(\alpha) b \quad (45)
$$

From (45), the biocomplexity of an organism is the product of its maximum rate of change of action $\dot{\Phi}(\alpha)$ measured in calories and its genome physical complexity (nonredundant functional genome size) measured in bytes. This again confirms what we said earlier—that the genome biocomplexity has informational as well as energetic components.

Definition of Life:

It is also appropriate to define life based on the first and second postulates of QIFFT: life is a quantum information fractal field that generates bioinformation oscillations through successive generations. Consequently, the first law of self-organization, equation (35), indicates that a system is said to be alive if:

$$\dot{\Phi}(0) > 0 \quad and \quad v(0) > 0 \quad (46)$$

Conversely, the system is dead if:

$$\Phi(0) = 0 \quad and \quad v(0) = 0 \quad (47)$$

Moreover, from equation (35), the following condition represents a system with constant stock of vitality (i.e., v(0) = constant):

$$\ddot{\Phi}(0) = K\dot{v}(0) = 0 \quad (48)$$

Like a virus, such a system does not metabolize and has weak stability analogous to that of undamped harmonic oscillator as is evident from equation (22). Nonetheless, a virus, or in general an RNA replicator, being a vitality system, can evolve, again based on equation (22), as we demonstrate in an upcoming chapter.

5.3—MAXIMUM ACTION MECHANISM

What support the claim that living systems are subject to a maximum-action principle, and what is the mechanism? We think there are two fundamental mechanisms that substantiate the principle of maximum action.

Kinetic power of replication:

This is based on the proposed "replication first" school of thought, which holds that a simple molecular replication reaction was the first step on the long road to life (Eigen 1992; Lifson 1997). The inspection of the characteristics of the primal replicating molecule may provide insights as to what is unique about the entire set of life chemical reactions and electrodynamics properties. Therefore, on the one hand, we regard the replication reaction, which being autocatalytic, has unique kinetic properties and constitutes the ultimate example of a kinetically driven reaction, (Pross 2003). On the other hand, we regard fractal phase conjugation, which generates maximum coherence and optimum distribution of charge due to golden ratio based replicator fractal geometry Winter (2012). Considering the kinetic power of autocatalytic replication reaction, Lifson (1997) has proposed the following numerical example in which there are two reactions:

$$A + B \rightarrow^X C \text{ (i)}$$

$$A + B \to^X X \text{ (ii)}$$

Reaction (i) is just a general representation of any chemical process—reactants A and B are converted into C through the catalytic effect of X. By contrast, reaction (ii), the molecular replication reaction, is an autocatalytic reaction in which the catalyst X converts A and B into more of itself. For reaction (i), if we assume a single molecule of catalyst X and an arbitrary reaction rate of 10^{-6} s/molecule, a period of 6×10^{17}s (derived from $6\times 10^{23} \times10^{-6}$s) or twenty billion years would be required in order to generate a mole of product C. On the other hand for reaction (ii), due to the enormous kinetic power of replication, it would take just a tiny fraction of a second for a mole of product X to be generated. The mathematics of replication is such that a single replicating molecule undergoing some seventy-nine acts of replication becomes a mole ($6\times 10^{23} \approx 2^{79}$), so in the above example, it would take just 79×10^{-6} s for a molecule to become a mole. The relative magnitude of these two figures, twenty billion years for catalytic reactivity and 79 μ s for autocatalytic reactivity, though dependent on the particular reaction parameters that are chosen, is striking and makes it clear that the enormous kinetic potential associated with the replication reaction places it in a unique kinetic category.

Thus the kinetic pathway leading to molecular replication is likely to be favored over any competing pathway, even if the rate of replication is many orders of magnitude slower than the rates of the competing routes. The autocatalytic replication reaction, by its very nature, is an extreme expression of kinetic control and will tend to overwhelm any competing reaction, thermodynamically preferred or not.

Consequently, Pross (2003) concludes that the kinetic power of autocatalysis effectively transcends thermodynamics, not through negation of the second law but by steering a kinetically driven and directed autocatalytic pathway that at all times remains fully consistent with the second law. Thus he considers this kinetic phenomenon an enormously powerful driving force (i.e., the driving force responsible for the emergence and evolution of life).

To the question regarding whether there is some physical principle that would anticipate a process in which a replicating entity becomes increasingly complex and metabolic, Pross suggests that within replicator space, the space in which dynamic kinetic stability (DKS) is effectively in control, the selection rule becomes kinetically less stable to kinetically more stable. However, he draws attention to the inherent difficulty in the formal quantification of DKS, which has manifestation at several levels. The difficulty arises from the fact that Pross has restricted his model to the chemical aspect of the replicator, whereas a more comprehensive approach necessitates the accommodation of the quantum electrodynamics aspect, which is the basis and complement to the chemical aspect. Within such a holistic approach, the driving force manifests itself through the correlation between the kinetic power of replication and fractal field phase conjugation that generates biocomplexity (DKS) or bioinformation (measured in terms of metabolism and information), as indicated by the first law of self-organization. So let us consider the quantum electrodynamics aspect of the replicator.

Fractal Field Phase Conjugation

The fundamental problem a living system has to resolve is how to gather or suck electric charge waves without destructive interference. The way the living system functions indicates that it has solved this problem in an extraordinary marvelous manner, while it should not from the perspective of quantum mechanics. From a quantum mechanical perspectives decoherence rather than coherence should prevail due to either environmental interaction or organism macroscopic localization. So how does the living system solve the problem? Whilst destructive interference is the norm in wave interference, the only exception in nature is when the waves interfere with golden ratio wavelengths (Winter 2012). Therefore, when waves of electrical charge are arranged in self-similar or fractal geometry optimized by golden ratio, they recursively and constructively interfere (heterodyne). The recursive constructive interference turns compression into acceleration, because golden ratio allows the wave velocities as well as the wavelengths to recursively add and multiply." We have been teaching for years that the only perfectly fractal three-dimensional electric field is the golden mean stellated dodecahedron. The beautiful thing is that in this structure, the best possible combination of wave interference occurs for a constructive output. This is because—as you well know—only the golden mean ratio allows constructive interference of both wave addition and wave multiplication" (Winter 2012).

Therefore, life phenomenon by selecting golden ratio based fractal dodecahedron, which is the geometry of DNA and proteins, it selects fractal field phase conjugation of maximum coherence and optimum distribution of charge. Consequently, it selects the path of maximum action.

Phase conjugation: when pairs of pinecones learn to kiss noses, by Dan Winter:

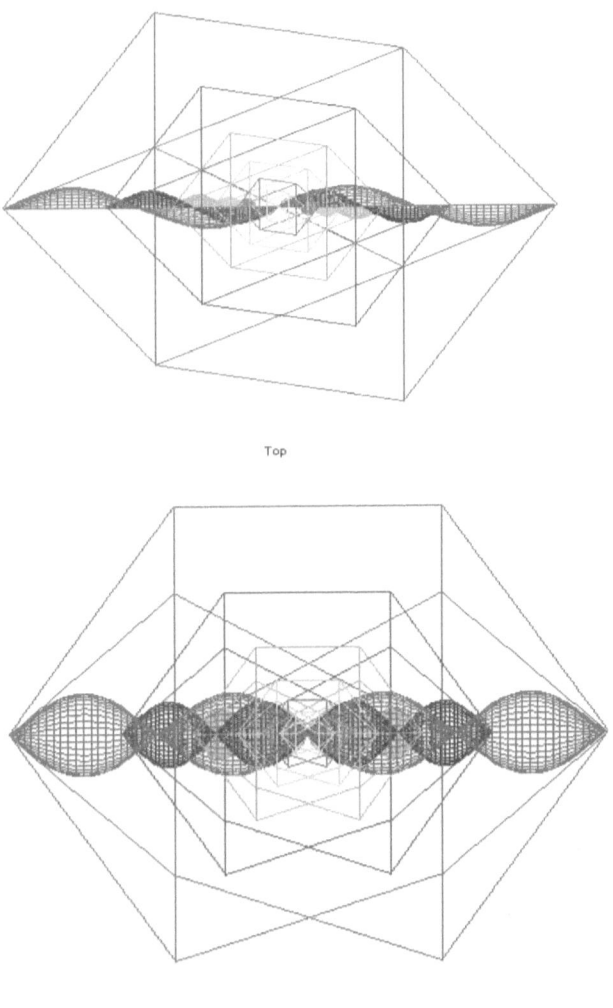

Top

Front

Source: http://www.fractalfield.com/mathematicsoffusion/

CHAPTER 6

EVOLUTION

6.1—BIOLOGICAL EVOLUTION GOAL FUNCTION

The life-organizing principle (20) describes the life process under constant environmental conditions. However, we know that environmental conditions do not always remain constant; they change. In doing so, the organism interacts and sometimes incorporates these changes and perturbations. The most significant perturbations from an evolutionary point of view are mutational changes. These perturbations induce nonlinearity in the system. Accordingly, the system may spiral to a focus due to positive damping, in which case the mutation is lethal and the system may become extinct. On the other hand, the system may exhibit a stable periodic solution in the neighborhood of the homogeneous solution for a greater period due to negative damping, in which case the mutation is beneficial. Thus we say that the system has evolved. The terms positive and negative damping are used in electrical engineering in connection with oscillatory phenomena that dissipate and absorb energy respectively. Likewise, we use these terms to characterize the dissipation and generation of vitality or bioinformation by the system when it interacts with the environment.

To incorporate evolution in our model, we make the following assumption:

Assumption 6.1:

Evolution (mutation, selection, and so forth) of a unicellular organism is a process through which the life-organizing principle undertakes negative damping.

Result 6.1:

The evolution of a unicellular organism leads to the increase of its total vitality, V(A).

Proof:

From (48), consider a virus with vitality v(0) = constant = $\overline{v}$, then using (22) we get:

$$\ddot{L} + \omega_1^2 L = 0 \quad (49)$$

therein, $w_1 = \dfrac{\overline{v}}{k} \quad (50)$

Now let the virus before evolution be given by (49) and assume it has evolved (i.e., equation (49) has undertaken negative damping). The new evolved system becomes:

$$\ddot{L} + \omega_2^2 L = \varepsilon f(L, \dot{L}) \quad (51)$$

Therein, ε is a small positive parameter that characterizes the smallness of the deviation of $\omega_2^2 L$ from $\omega_1^2 L$. If the deviation is not small, we set $\varepsilon = 1$. Biologically, $\varepsilon \neq 0$ accounts for certain generation of information that did not exist when $\varepsilon = 0$. Being interested in finding a stable periodic solution in the neighborhood of the homogenous solution, under the condition of negative damping, equation (51) is readily solvable by Krylov and Bogoliubov (1947) and their asymptotic method for damped oscillations. The solution yields:

$$\omega_2 < \omega_1 \quad (52)$$

Given, $V(A) = \bar{v}A$, $w = \dfrac{\bar{v}}{k}$, $w = \dfrac{n\pi}{A} = n\pi f$, $k = bGA^a$,

Note: $\omega = n\pi f$, n = 1, 2, ..., $\pi = \dfrac{22}{7}$,

This follows:

$$V(A) = \frac{n\pi bG}{f^a} \alpha \frac{1}{\omega^a} \quad (53)$$

Then:

$$V_1(A) \ \alpha \ \frac{1}{\omega^a_1}, \quad V_2(A) \ \alpha \ \frac{1}{\omega^a_2} \quad (54)$$

$V_1(A)$, $V_2(A)$ is the total vitality of the organism before and after evolution, respectively. Equations (52) and (54) yield:

$$V_2(A) > V_1(A) \quad (55)$$

This proves our result.

Equation (55) proves that the evolution of the unicellular organism leads to the increase of its total vitality. However, we need to know what the increase of total vitality means.

For now, referring to figure 1, the increase of total vitality (the area under the vitality curve) leads to the increase of at least one of the following:-

[A] Life span or cell cycle time, A

[B] Bioinformation or biocomplexity, $v(\alpha)$

[C] Matter-energy growth function (i.e., body size or mass)

Later on, we derive a specific equation for the increase of total vitality. It is appropriate to note that researchers used to consider one parameter or factor, such as body size or life span or biocomplexity, as a measure of evolutionary progress; then they usually confronted or were challenged by exceptions and hence concluded that evolution is most probably random. However, it is clear from (55) that evolution as an increase of total vitality is not defined by one parameter (e.g., biocomplexity, life span, or body size); rather, it is a product of all these independent factors, which then facilitate a comprehensive evolution goal function or target criterion upon which natural selection acts.

Note: If we redefine $\bar{v}$ to be average vitality density for all ages:

$$\bar{v} = \frac{V(A)}{A}$$ = constant, then again (22) can reduce to the simple form (49), solvable by the Krylov and Bogoliubov method, in order to account for metabolic unicellular organisms or multicellular organisms.

6.2—SECOND LAW OF SELF-ORGANIZATION

As we have seen in (6.1), assuming evolution (mutation, selection, and so on), a process through which the life-organizing principle undertakes negative damping, proves that evolution leads to the increase or maximization of total vitality, which is the area under the vitality curve, V(A). The maximization of total vitality as goal function solves

Dobzhanesky's fundamental problem: "The origin of organic adaptedness, or internal teleology, is a fundamental, if not the most fundamental, problem of biology" (Dobzhansky et al. 1977, 95).

It is also significant to show that bioinformation (total vitality) is also quantized. Such quantization has its expression in beneficial mutational changes that uncover the genome quantum stable or quasi-stable states. This means that although mutational changes are random, beneficial mutation states are not random; they have to concord with DNA quantum functionally stationary states. For this purpose, the life-organizing principle could be employed to derive the proposed quantization relationship. The derived quantization relation could provide plausible theoretical basis for punctuated equilibrium. Accordingly, we set the boundary conditions for the life-organizing principle according to the fact that an organism is trapped within its own life span; trapped within an irreversible time domain, it lives neither before birth nor after death. Thus given (20) and (14), we get:

$$L(t) = L_0 e^{\frac{iV(t)}{k}} = C\cos\frac{V(t)}{k} + B\sin\frac{V(t)}{k} \quad (56)$$

We assume that the time, t, which is a measure of an organism age, is at the same time a measure of bioirreversibility, the main difference between life and nonlife (i.e., the bioirreversibility of an inanimate system is t = 0).

Setting the boundary conditions:

At t = 0 biotic irreversibility is zero, so V (0) = 0, and we let L (0) = 0.

At t = A the system is dead and v(A) = 0, so we let L(A) = 0.

$$\therefore 0 \leq t \leq A \quad (57)$$

This means that the life of an organism is bounded (closed within the interval (57)), so it has no life before birth and has no life after death.

From (56) and the boundary conditions, we get:

$$C = 0, \quad \frac{V(A)}{k} = n\pi, \ n = 1, 2, ..$$

$$\therefore V(A) = n\pi k = n\pi \, bGA^a \quad (58)$$

$$= \frac{n\pi \, bG}{f^a} \quad (59)$$

Note that the life span A is at the same time the period of oscillations. Equation (59) is the second law of self-organization; it is a quantum fractal law of evolution and development. It is a power law, a Pareto (1897) type of law, in which total vitality is inversely proportional to the frequency of self-sustained bioinformation oscillations. It is scale invariant; it describes the development and evolution of genes, cellular organisms, multicellular organisms, and ecosystems, reflecting the self-similarity of biological hierarchy. It is interesting that Bak and Paczuski (1995) have also proposed a power law to account for punctuated equilibrium (N. Eldredge and J. S. Gould). A shortcoming of the law they propose is that it is not derived from a theory of general biology that is, it is abstract. Moreover, the dependent variable (imagined accumulated biological quantity) varies inversely with time (pure or absolute time), so the criticality is not imposed by the system's internal dynamics.

Elsheikh (2010) indicates that speciation is a punctuationistic transition from a lower major attractor (life-organizing principle) to an upper major attractor, governed by the second law of self-organization. Consequently, the existence of disjoint stationary states (attractors) implies that transitional forms are genomically unstable, transient dwellers, which may explain the absence or scarcity of their fossil records.

The constant G could be shown to represent the organism total action; hence it depends on the species. Consider equation (14) and equation (18):

$$\frac{V(t)}{\Phi(t)} = \frac{b \int_0^t E(x)(A-x)^a dx}{A^{-a} \int_0^t E(x)(A-x)^a dx} = bA^a \quad (60)$$

$$\therefore V(t) = bA^a\Phi(t)$$
$$\therefore V(A) = bA^a\Phi(A) \quad (61)$$

From (58) and (61), we get:
$$n\pi GbA^a = bA^a\Phi(A)$$
$$\therefore \Phi(A) = n\pi G \quad (62)$$

Equation (61) shows that evolution of an organism, being maximization of total vitality, does not only maximize genome physical complexity (b) and life span (A) but also maximizes total action ($\Phi(A)$). It seems evident—based on the maximum action principle—that these three factors facilitate an exclusive evolutionary goal function and fitness unit upon which natural selection acts (i.e., natural selection selects the path of maximum action). Consequently, the most evolved organism is the one that has the greatest product of these factors as given by equation (61).

The Second Law of self-organization, equation (59), is both fractal and quantized. We show that both aspects are related to Fibonacci numbers or the golden ratio. This can be revealed by inquiring about the nature of the integer n. The integer n is supposed to account for the functional stationary quantum states of a living system as well as the stability and functionality of its pattern formations. For this sake, let us suppose that an organism having total action ϕ_1 produces another organism of total action ϕ_2. As a consequence, three possibilities emerge:

Case1— $\phi_2 = \phi_1$: In this case, organism ϕ_1 is said to be stable or in a stationary state.

Case2— $\phi_2 < \phi_1$: In this case, organism ϕ_1 is said to be degraded. It is a path favorable by the second law of thermodynamics and the least-action principle.

Case3— $\phi_2 > \phi_1$: In this case, the organism ϕ_1 is said to grow or evolve, i.e., manifests the maximum action principle. However, adequate representation of the maximum action principle necessitates the following conditions:

i) $\phi_2 > \phi_1$.

ii)   The second law, being fractal, or power law, $\dfrac{\phi_2}{\phi_1}$ must be equal to $(\phi_1 + \phi_2)/\phi_2$ to preserve the scale invariance. Moreover, this condition emphasizes that the maximum action principle forces the maximum value of $\dfrac{\phi_2}{\phi_1}$.

Since from equation (62), ϕ is directly proportional to n, where n is an integer, then these conditions can be restated as follows: $n_2 > n_1$ and n_2/n_1 is maximum. It follows that to satisfy conditions i) and ii), we must have:

$$\frac{n_2}{n_1} = \frac{n_1 + n_2}{n_2} = \frac{n_1}{n_2} + 1 \text{ , let } \frac{n_2}{n_1} = x \text{ we get:}$$

$$x = \frac{1}{x} + 1 \quad \therefore x^2 - x - 1 = 0$$

$$\therefore x = \frac{n_2}{n_1} = \frac{1 + \sqrt{1+4}}{2} = 1.618.. \text{ = golden ratio}$$

This means that the golden ratio is a deterministic consequence of the maximum action principle and that n in the fractal law, equation (59), is Fibonacci's number. Thus the second law, $V(A) = \dfrac{n\pi \overline{G}}{f^a}$, ($\overline{G} = bG$), is fractal—being a power law—and that total vitality is quantized in accordance with Fibonacci numbers. This result emphasizes that the functional stationary quantum states of a living system, being golden ratio fractal, is in agreement with Winter's (2012) discovery that biofield fractal phase conjugation based on golden ratio provides optimum coherence, hence optimum functionality.

Finally, applying the second law to gene evolution (or RNA evolution) indicates that the QIFF, being a self-organizing field, structures nucleotides along a path of maximum action and maximum bioinformation (i.e., a path that maintains synthesis of viable functional proteins and generates self-sustained bioinformation oscillations). It is a path of total vitality increase that is in direct correspondence with Fibonacci numbers and nucleotides physical complexity, and it varies inversely with the frequency of nucleotides replication cycle time. It is a path of syntactically meaningful genetic code.

6.3—PUNCTUATED EQUILIBRIUM

From (59), the incremental increase in total vitality is given by:

$$\Delta V(A)_i = \pi \left(\frac{m\overline{G}_m}{f_m{}^a} - \frac{n\overline{G}_n}{f_n{}^a} \right) \Rightarrow L_i(v) \quad (63)$$

Herein, m and n are Fibonacci numbers where n > m, and i = 1, 2, 3 . . . is an integer. The Fibonacci numbers characterize the quantum functional stationary states, while $L_i(v)$ represents the number of minor attractors generated in case of multicellular organisms evolution. Note: due to Fibonacci sequence properties, the addition or subtraction of the consecutive life-organizing principle attractors also produces attractors.

According to (63) the incremental increase of total vitality (the beneficial mutational changes) in the early stages of speciation (that is, at low frequencies) is large compared to the incremental increase at high frequencies. Therefore, in the process, stasis could be maintained within a given niche configuration. It follows that the DNA (species) evolves in a lawful manner of discrete probabilistic nature to maintain higher stable states of total vitality. It is clear that equation (63) has a punctuationistic behavior, meaning that rapid evolutionary changes in the initial stages that fall down into stasis (Eldredge and Gould 1972; Gould 1982). In general, we envisage (59) as the law which describes phylogenic evolution. Therefore, we make a distinction between microevolutionary changes and macroevolutionary ones.

Microevolutionary change is little improvements (minor increases in total vitality) within the same major attractor, whereas macroevolutionary change is a transition from one major attractor to another one. The unit for such transition is a minor attractor; a cell type is an example or representative of a minor attractor. Consequently, macroevolution is the generation and assembly of minor attractors. This view is in accord with Root-Bernstein and Dillon's (1997) theory of molecular complementarity, which ascertains that biotic evolution occurs by assembling compact subassemblies. By complementarity, they mean nonrandom reversible coupling of the components of a system. However, the notion of compact subassembly lacks both rigorous quantitative definition and dimension.

It is also clear from (59) that a life or bioinformation oscillation is a low-frequency phenomenon (i.e., macroscopic phenomenon). The period of oscillations is given in minutes, hours, or years. This supports our proposed generalized complementarity: a material system does not possess simultaneously both matter waves and information oscillations for its description. Although total vitality quantization may differ from energy quantization in the details, such as the nature of boundary conditions, the essence remains the same (i.e., nature is discrete). In fact, the existence of evolutionary disjoint functionally stationary states (major attractors) imposes the discreteness of total vitality.

To provide a more realistic picture of the evolutionary process and to account for some quantitative features of population and organism growth dynamics, we consider, from equation (16), the genome total bioinformation, which is given by:

$$T(A) = V(A) + F(N(A))$$

V(A) is total vitality; F(N(A)) is total natality density function.

Now assume under certain evolutionary and environmental perturbations that the genome's total bioinformation is conserved phylogenetically along a certain lineage:

$$T(t) = U(t) + Z(N(t)) = constant \quad (64)$$

U(t) is the species or population mean total vitality at time t during successive generations, and Z(N(t)) is total natality density function at time t during successive generations.

This assumption emphasizes the stability of the genome, which preserves the species properties and indicates that it is in a stationary state. Moreover, the assumption offers the following desirable results:

A. Since, in general, the evolutionary process maximizes total vitality, so—based on (64), evolution also minimizes natality. This resolves Waddington's problem: "For us, the major problem is one which was only a second order issue for Darwin. This is the problem of adaptation. Why do we find animals and plants which have structures and capacities that make them admirably suited to carry out extraordinary living routines in the most unlikely situations, often highly unfavorable for reproduction?" Waddington (1968).

B. If the population or the species is somehow forced to regress (that is, its total vitality decreases), then from (64), its natality rate must increase. Otherwise, the species is susceptible to extinction.

The problem with neo-Darwinism is that it restricts survival to reproductive fitness, whereas, according to (64), the genome's total bioinformation generates two survival components: reproductive fitness and total vitality (fitness). Evidently, survival is a product of the whole genome's function. In this context, the conservation of genome's total bioinformation accounts for the recent developments in sociobiology concerning multilevel selection. It is found that there is lower-level selection (individual selection) as well as higher-level selection (group selection), and that higher-level selection (whose fitness unit we now regard to be total vitality) takes place at the expense of individual reproductive fitness (Wilson 2007). On this basis, it is appealing to argue that the evolution of major transitions is also subject to the conservation of total bioinformation.

6.4—LIFE BEFORE EARTH

Alexei Sharov, from the National Institute on Aging in Baltimore, and Richard Gordon, from the Gulf Specimen Marine Laboratory in Florida, published a paper in the online journal ArXiv (2006). Using Adami's (2002) definition of genome's physical complexity, they indicate that if we plot genome complexity of major phylogenetic lineages on a logarithmic scale against the time of origin, the points appear to fit well to a straight line (Sharov 2006), figure 3. Genome complexity increased exponentially and doubled about every 376 million years, and they concluded the following: "Linear regression of genetic complexity (on a log scale) extrapolated back to just one base pair suggests the time of the origin of life = 9.7 ± 2.5 billion years ago." Since the age of Earth is only 4.5 billion years, life could not have originated on Earth, contrary to the widely accepted view.

To achieve this result, Sharov and Gordon were inspired by Moore's law, according to which the complexity of computers grows at a rate of double the transistors per circuit every two years, resulting in exponential growth. Looking at the complexity of computers today and working Moore's law backward shows that the first microchips came about during the 1960s, which is when they were actually invented. In their paper, Gordon and Sharov take the same approach, only they apply it to biocomplexity. They suggest that the complexity of life and the rate at which it has increased follows Moore's law, but in this case, the doubling time is 376 million years rather than two years.

Sharov and Gordon reject the idea that perhaps the early steps in the origin of life created complexity much more quickly than evolution does now, which will allow the timescale to be squeezed into the life span of the Earth. In fact, this is exactly what Elsheikh (2010, 2013) tried to prove: biological evolution, driven by the maximum action principle, has taken the least time from the origin of life up to now. This may be evident if we compare the space of functional proteins with the space of all possible proteins (Eden 1966).

To clarify this point of view and make a correction to the Sharov and Gordon model, it is important to distinguish between physical information and bioinformation. Although Adami (2002) argues correctly that correlations within a sequence are not going to reveal how the sequence is correlated to the environment within which it is to be interpreted. So that genome's physical complexity must be reflected in the structural complexity of the organism that harbors it. Nonetheless, his definition of physical complexity as nonredundant functional genome size does not correspond to organism complexity (biocomplexity), which is developmental and functional. In other words, although Adami assumes that genome's information is about the organism and it is functional, unfortunately his definition does not reflect his intention. For the definition to account for bioinformation as developmental functional complexity (organism), it has to have the dimensions of energy and information, since there is no biological function without energy. This necessitates a distinction between genome physical complexity, which can be measured in terms of nonredundant functional size (bytes), and genome biocomplexity, which can be measured in terms of nonredundant functional genome size and energy (bytes x calories). Using equation (45), which defines biocomplexity for different species, we get:

$$v_s(\alpha) = \dot{\Phi}(\alpha)b$$

Where $v_s(\alpha)$ is genome biocomplexity, $\dot{\Phi}(\alpha)$ is the organism maximum rate of change of action (in calories) that takes place at adulthood and b is nonredundant functional genome size in (bytes). This equation is different from that given by Sharov and Gordon only by $\dot{\Phi}(\alpha)$, which represents the organism's metabolic rate at adulthood (i.e., when fully grown). It is appropriate to mention that Ulanowicz (2004) revealed that biological organization has the dimensions of energy and information and is measurable in terms of ascendency units (flow x bits).

Now it is mathematically clear that if we plot $v_s(\alpha)$ versus time of origin in the same manner that Sharov and Gordon did, we are going to have a slope with a greater angle than Sharov and Gordon achieved. Thus the extrapolation of the new straight line most probably will allow the timescale to be squeezed into the life span of the Earth. The timescale predicted by Sharov and Gordon is the time for the evolution of genome complexity had there been no maximum action principle. According to the maximum action principle, the rate of action increase is proportional to vitality, which is a measure of the organism's developmental functional complexity (bioinformation). Therefore, since vitality traverses a path of maximum action and since evolution maximizes total vitality, it follows that natural selection selects the path of maximum action. Consequently, it is conceivable to envisage life as having originated on Earth about 3.8 billion years ago instead of Sharov's 10 billion years ago.

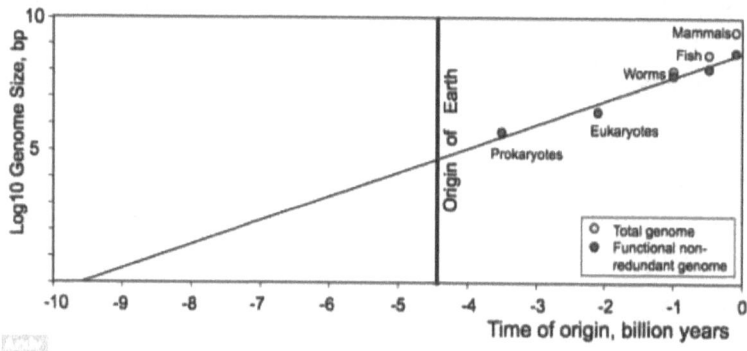

Fig 3. On this semilog plot, the complexity of organisms, as measured by the length of functional nonredundant DNA per genome counted by nucleotide base pairs (bp), increases linearly with time (Sharov 2012). Time is counted backward in billions of years before the present (time 0). Modified from figure 1 in (Sharov 2006).

Source: http://phys.org/news/2013-04-law-life-began-earth.html#jCp

CHAPTER 7

GROWTH AND DEVELOPMENT

7.1—POPULATION GROWTH:

Populations of organisms usually increase as much as they can in their environments. However, populations do not grow forever; some form of resistance from an environment will stop their growth. The form of environmental resistance that limits population growth is called a limiting factor. These changes in population can be graphed, usually in accord with one of two graphs: the J-curve graph or the sigmoid curve (S-curve). Initially, both graphs have similar growth, where there are very few organisms in the beginning so the reproduction rate is low and the population grows very slowly. Then the population begins to grow very quickly in an exponential manner. However, the population cannot maintain the exponential growth for long due to limiting factors. Limiting factors that produce the J-curve are density independent, which means that they can affect a population at any density. These are natural disasters such as fire, flood, drought, temperature, weather, sunlight, and so forth. Usually, density independent limiting factors cause a sudden crash or decline in population.

Limiting factors that produce an S-curve are density dependent, which means that they only affect a population when it reaches a certain density. These are competition, predation, and disease, which usually affect a population at certain densities. The S-curve necessitates an environment that is able to renew or recycle resources continuously and that the population's density is somehow can be decreased and enters a deceleration phase. Then the population enters an equilibrium phase, where the number of births equals the number of deaths in a population. Usually, the population fluctuates around the carrying capacity.

Why do different species have the same population and organism growth patterns, i.e., the empirical logistic curve or J-curve? Could it be a reflection of an underlying principle? Yes, we think so; it is a reflection of conservation of the genome's total bioinformation. It follows that total bioinformation conservation can usefully be employed to generate population and organism growth dynamics. Thus we use equation (64):

Given T(t) = U(t) + Z(N(t)) = constant

$$\therefore \ \dot{T}(t) \ = \ \dot{U}(t) \ + \ Z'_N \dot{N} = 0$$

$$\therefore \ N(t) = \ c \ - \ \int_0^t \frac{\dot{U} \, dx}{Z'_N}, \quad (65)$$

Where $Z'_N = \dfrac{\partial Z}{\partial N} \neq 0$, and c is a constant of integration.

$$\therefore \ \frac{dp}{dt} \ = \ p(c \ - \ \int_0^t \frac{\dot{U} \, dx}{Z'_N}) \qquad (66)$$

    a)   During the initial growth phase, environmental pressure being little, it follows:

$$\dot{U} \ = 0 \ \Rightarrow \ p_t \ = \ p_0 e^{ct} \quad (67)$$

This is a phase of exponential growth.

    b)   When environmental pressure increases and density dependent factors come into effect, then:

$$\dot{U} < 0 \, , \quad Z'_N < 0 \quad \text{, and it follows that a time will}$$

come when,

$$\Delta \; = \; c \; - \int_0^t \frac{\dot{U}\,dx}{Z'_N} \; = \; 0 \qquad (68)$$

$$\therefore \; p_t \; = \; p_{max} \; = \; \text{constant} \quad (69)$$

Therefore, population growth dynamics is subject to the conservation of the genome's total bioinformation.

7.2—ORGANISM GROWTH:

The development of a multicellular organism from a single cell is a spectacular form of self-organization. Although nearly all cells within an organism are genetically identical (Albert 2002), internal processes lead to the emergence of highly organized structures such as skins, eyes, hearts, and brains. Organism growth is achieved by an increase in cell size, an increase in cell number, and by deposition of extracellular materials such as bone and shell. This process is in concert with pattern formation and cell differentiation. Cell differentiation is the process that renders cells structurally and functionally different from each other, leading to distinct cell types such as muscle or skin cells.

How growth, pattern formation, and cell differentiation are coordinated during development is largely unknown. However, there is evidence that special signaling molecules, which are known as morphogens, play a key role in the coordination of these processes. Nonetheless, little is known about the principles of growth control; how the sizes of animals or organs are determined still remains an open question (Conlon 1999). For the normal growth of organs during development, extrinsic factors such as nutrients and hormones are essential. However, these extrinsic factors do not determine the size under normal conditions. Instead, it has been suggested that the size of organs is determined by organ-intrinsic mechanisms (Bryant 1984).

To tackle the problem of organism growth from the perspective of QIFFT, two principles are to be considered:

1- Major attractor principle: According to this principle, each multicellular organism is composed of a major attractor and minor attractors. The major attractor is the life-organizing principle that represents the organism (as a whole) dynamics for successive generations. The minor attractors are the life-organizing principles representing the different cell types. We propose that the minor attractors belong to the basin of the major attractor. Consequently, the minor attractors are constrained to superpose and cohere to concord with the major attractor. This is how the genome, or QIFF, or life-organizing principle exercises control on an ontogenetic developmental path, which is a path of maximum action. If this hypothesis is correct, then probably the minor attractors represent a Fourier expansion of the major attractor. This is so ontogeny recapitulates phylogenetic evolution in a new perspective; both processes generate and assemble minor attractors in accordance with the second law of self-organization.

2- Conservation of genome's total bioinformation: This principle is applicable to the major attractor as well as to the minor attractors; in both cases, it determines the final size of the growing system, as shown above with respect to population growth. In the same manner, the principle determines the final size of a cell type population (e.g., tissue or organ, governed or generated by a minor attractor). Note that both the cell types and the organism as a whole are subject to the second law of self-organization. Ontogenetically, cellular differentiation, which generates cell types, is a process through which the life-organizing principle (minor attractor) undertakes negative damping, leading to an increase of total vitality in accordance with the second law of self-organization, whereby Fibonacci numbers characterize the cell type quantum stationary functional state.

Now, because every multicellular organism grows through successive divisions of one cell (i.e., zygote), and because successive divisions of the zygote can be described by the life-organizing principle, we account for organism growth by considering the following assumptions:

Assumption 1—Cellular differentiation is a process through which the life-organizing principle undertakes negative damping. Particularly, it leads to the increase of cell cycle time (Shackney 1973; Voit 1984).

Assumption 2—The genome total bioinformation is conserved through progressive cellular differentiation. Cloning may be a good support for this assumption.

Based on these assumptions, equation (66) is applicable to organism growth. Where p(t) is now the number of cells and U(t) is now the organism's cells mean total vitality at time, t. It follows at:

Cleavage phase: in this phase, $\dot{U} = 0$, $\Rightarrow p_t = p_0 e^{ct}$ (70)

Cellular differentiation phase: In this phase, $\dot{U} > 0$, $Z'_N > 0 \Rightarrow$, and then a time will come when (68) holds, then:

$p_t = p_{max} = $ constant (71)

Limited growth (Zotin 2006) is maintained, using (68), for:

$$\Delta = c - \int_0^t \frac{\dot{U}\,dx}{Z'_N} < 0 \quad (72)$$

The relationship between phylogenic evolution, which leads through progressive increase of total vitality (specialized organisms) to stable ecosystems and ontogenetic development, which leads through progressive increase of total vitality (specialized cells) to stable organisms is apparent: both phenomena are expressions of the conservation of genome's total bioinformation.

7.3—MONTAGNIER'S REVOLUTIONARY DISCOVERY

Luc Antoine Montagnier is a French virologist and joint recipient with Françoise Barré-Sinoussi and Harald zur Hausen of the 2008 Nobel Prize in Physiology or Medicine for his discovery of the human immunodeficiency virus (HIV). Furthermore, Montagnier (2009, 2010) has brought forth remarkable evidence for a nonparticle view of life. He claims that DNA can send electromagnetic imprints of itself into distant cells and fluids, which can then be used by enzymes to create copies of the original DNA. The basic setup of his experiments was that two adjacent but physically separate test tubes were placed within a copper coil and subjected to a weak extremely low frequency electromagnetic field of seven hertz. The apparatus was isolated from Earth's natural magnetic field to stop it from interfering with the experiment. One tube was thoroughly filtered from a fragment of DNA around one hundred bases long; the second tube contained pure water. After sixteen to eighteen hours, both samples were independently subjected to the polymerase chain reaction (PCR), a method routinely used to amplify traces of DNA by using enzymes to make many copies of the original material. The gene fragment was apparently recovered from both tubes, even though one should have contained just water. Thus it would be possible, according to Montaigner's results, to duplicate viral and bacterial DNA in the absence of the physical template of the DNA itself. Coding information would be transmitted by electromagnetic waves generating from water molecules.

Physicists in Montagnier's team (2010) suggest that DNA emits low-frequency electromagnetic waves that imprint the structure of the molecule onto the water. This structure, they claim, is preserved and amplified through quantum coherence effects, and because it mimics the shape of the original DNA, the enzymes in the PCR process mistake it for DNA itself and somehow use it as a template to make DNA matching that which "sent" the signal.

Many scientists greeted Montagnier's claims with scorn and harsh criticism. One of the criticisms of the work was that there is no known mechanism by which bacteria can generate radio waves. In addition, experiments carried out by Montagnier raise the fundamental question of how water could store and receive electromagnetic information of such precision that a DNA sequence could be reproduced without a template, which is how it is normally done. Jeff Reimers is a theoretical chemist at the University of Sydney, Australia, points out: "If the results are correct, these would be the most significant experiments performed in the past 90 years, demanding re-evaluation of the whole conceptual framework of modern chemistry."

To account for Montagnier's revolutionary discovery, a quantum biological revolution is also needed. It is clear that the discovery challenges mainstream biology in that the principle that all life comes from life could hold only on the basis of a nonparticle view of life. Moreover, the discovery being about encoding, transmission, and decoding of bioinformation lies beyond the boundaries of ordinary quantum field theory. To address this fundamental problem, Elsheikh (2010; 2013) proposes a new quantum information fractal field theory. The quantum information fractal field, representing the DNA or genome, is a self-organizing field that structures nucleotides along a path of maximum action and maximum bioinformation, i.e., a path that maintains synthesis of viable functional proteins and generates self-sustained bioinformation oscillations. It is a path of syntactically meaningful genetic code. According to the maximum action principle, the rate of change of action (matter-energy metabolism) is proportional to the genome capacity to generate developmental functional complexity (vitality or bioinformation). Moreover, the path of maximum action, which is in fact a path of total vitality, increase is characterized by the increase of Fibonacci numbers, number of nucleotides (genome physical complexity), and varies inversely with the frequency of bioinformation oscillations. Vitality, a measure of bioinformation, has the dimensions of information and energy. Fibonacci numbers, being a deterministic consequence of the maximum action principle, characterize the quantum functionally stationary states (Elsheikh 2013).

Nucleotides sequence being a sequence of dodecahedrons, as White (2008) asserts, the path of maximum action and bioinformation is also a path of golden ratio optimized fractal dodecahedrons. Winter (2012) revealed that the electric geometry of DNA is the fractal nesting of golden ratio based dodecahedron. Within such geometry, electrical waves of charge nest in self-similar or recursive fractal like embedding. It follows that wave heterodynes recursively allow the wave velocities, adding and multiplying constructively. It is also called fractal field phase conjugation. This recursive heterodyning or fractal field phase conjugation allows charge to be accelerated coherently. Consequently, the generation of electromagnetic waves or signals (EMS) is associated with the acceleration of charges within a certain stationary functional quantum state, which is similarly a golden ratio optimized fractal dodeca quantum state.

Thus the EMS encode the information about the golden ratio optimized fractal dodeca quantum states. For water to be capable of storing this bioinformation, the EMS must reassemble and organize water molecules in sequence of golden ratio optimized fractal dodecas. Do water molecules have the property of being organized to form fractal dodecas? The answer is yes. Not only that, but also it is established that the dodeca is relatively stable and a common motif in water clusters (Loboda 2010). Now the induced EMS of bioinformation by creating golden ratio optimized fractal dodecas in water, they actually create DNA quantum information fractal field template (i.e., create a path of maximum action and bioinformation). The template is therefore capable of emitting the same EMS of bioinformation due to water dodeca fractal field phase conjugation; then given the essential DNA ingredients, the template can reassemble the whole DNA sequence. This indicates that fractal field phase conjugation is the mechanism by which both DNA and water clusters generate EMS of bioinformation. It is important to note that Winter (2012) discovered fractal field phase conjugation and made numerous technological inventions on this basis. In sum, the quantum information fractal field generates two types of oscillations: bioinformation oscillations that contain the dynamical essence of a living system and EMS, or waves, that encode the bioinformation.

Finally, it is appropriate to highlight some of the significance of Montagnier's discovery for medicine. Montagnier et al. (2010) indicates that the signals detected appear to be a property of most bacteria infecting humans, as well as many viruses, including HIV, influenza A, and hepatitis C. Further, it appears from the research that some common diseases not previously considered to be of bacterial origin may indeed be so. In evidence of that, signals identical to those detected in test tubes containing live bacteria have been found in the blood plasma, and in the DNA extracted from the plasma, in patients suffering from Alzheimer's, Parkinson's disease, multiple sclerosis, chronic Lyme syndrome, rheumatoid arthritis, and various neuropathies. Montagnier has proposed to employ these radio frequency techniques for detection of chronic bacterial and viral infections and to explore means to use them in treatment of diseases, including AIDS and autism. Montagnier also notes that such techniques might someday provide a solution to the growing problem of evolution of antibiotic-resistant organisms.

CHAPTER 8

ECOSYSTEM DYNAMICS

8.1—ECOSYSTEM GOAL FUNCTIONS

Ecosystem dynamics is the dream habitat where the maximum action principle is dwelling. The habitat is occupied by numerous maximization theories: maximization of power (Odum 1983), maximization of ascendency (Ulanowicz 1986 and 2004) and maximization of eco-exergy (Jorgensen 2006 and 2007). Jorgensen emphasized that for the description of the dynamics of ecosystems, the three formulations are equally valid because they cover simply different aspects of ecosystems that are extremely complex, with extremely complex dynamics. It has been shown (Fath et al. 2004; Jorgensen 2007) that to maximize eco-exergy storage means also that the power and the ascendency are maximized.

It is interesting that the ecological law of thermodynamics (Jorgensen 2006) correlates the two types of theories: those which maximize the flux of useful energy and those which maximize biological organization and function. The law states: "Ecosystem development in all phases will move away from thermodynamic equilibrium and has the propensity to select the components and the organization that yields the highest flux of useful energy throughout the system and the most exergy stored in the system. This also corresponds to the highest ascendancy" (Jorgensen 2006).

According to Fath et al. (2004), ecosystem growth is the increase of energy through flow and stored biomass, and ecosystem development is the internal reorganization of these energy mass stores, which affect transfers, transformations, and time lags within the system. Several hypotheses thermodynamically describe the natural tendency that ecosystems follow during succession. Yet Fath et al. view ecosystem succession as a series of four growth and development stages: boundary, structural, network, and informational. They explain these stages by indicating that boundary growth is required as the system's initial input but continues throughout as the source of low-entropy energy for growth and to maintain order in the ecosystem. Structural growth is dominant during early succession, when stored exergy increases due to more biomass and more physical structure. During the third and fourth development stages, larger, longer lived, and genetically more complex plants and animals tend to replace smaller, shorter lived, and simpler ones. Clearly, these stages represent eco-exergy storage maximization as well as total vitality maximization.

Hence, to shed more light upon these developments and to suggest mathematical formulation for the ecological law of thermodynamics based on our proposed maximum action principle, we start by readapting the above-mentioned equations to ecosystem dynamics. This necessitates the following assumption.

Assumption 8.1:

Ecosystem dynamics reflects the underlying average dynamics of the constituent species.

According to this assumption ecosystem dynamics could be represented by the average dynamics of the ecosystem's species. At a glance, this may look reductionistic because the ecosystem is greater than the arithmetical sum of its constituents. However, due to our proposed notion of bioinformation, which ascertains that the bioinformation is about the phenotype (i.e., about representing the genome within specific environmental conditions), it follows that the species interactions with the environment, and the effect of such interactions, are incorporated within the species vitality. To some extent, this raises the dynamics to the level of species-environment interactions, which is a basic requirement of ecosystem dynamics. Consequently, the above-mentioned equations that deal with individual species take the following form in order to account for ecosystems:

Definitions:

Ecosystem vitality: $\overline{v}(t) = \dfrac{\sum\limits_{i=1}^{p} v_i(t)}{p}$ (73)

It measures the ecosystem developmental functional complexity at time t; p is the number of species.

Ecosystem genome complexity: $\overline{b}(t) = \dfrac{\sum\limits_{i=1}^{p} b_i(t)}{p}$ (74)

Ecosystem instantaneous life span: $\overline{A}(t) = \dfrac{\sum\limits_{i=1}^{p} A_i(t)}{p}$ (75)

$\therefore \overline{K} = \dfrac{1}{\overline{b}\,\overline{A}^{a}}$ (76)

Ecosystem life span: $A(\alpha) = \alpha$

Therein, α is the time when the ecosystem is fully mature. Notice that since an ecosystem, in general, does not die or divide like a unicellular organism, we limit our consideration to the time when it is fully mature: $A(\alpha) = \alpha$. This does not exclude the other option to look for a reasonable estimation for its life span.

Ecosystem instantaneous total vitality:

$$U(t) = \int_0^t \bar{v}(x)dx \quad (77)$$

Vitality, which is a measure of the ecosystem developmental functional complexity:

$$\dot{U}(t) = \bar{v}(t) \quad (78)$$

Ecosystem maximum average total vitality:

$$\bar{U}(A) = \frac{\sum_i^p V(A)_i}{p} \quad (79)$$

The phase of the ecosystem that represents the maximum action principle is given by:

$$\Psi(t) = \sum_{i=1}^p \Phi_i(t)/p = \bar{K} \int_0^t \bar{v}(x)dx \quad (80)$$

Relationship between ecosystem rate of change of action and its developmental functional complexity:

$$\dot{\Psi}(t) = \bar{K}\bar{v}(t) \quad (81)$$

Ecosystem maximum rate of change of action:

$$\dot{\Psi}(\alpha) = \bar{K}\bar{v}(\alpha) \quad (82)$$

Ecosystem biocomplexity:

$$\bar{v}_s = \frac{1}{\bar{A}^a}\bar{v}(\alpha) = \frac{1}{\bar{A}^a}\frac{\dot{\Psi}(\alpha)}{\bar{K}} = \bar{b}(\alpha)\dot{\Psi}(\alpha) \quad (83)$$

Ecosystem total action:

$$\Psi(\alpha) = \bar{K} \int_0^{\alpha} \bar{v}(t)dt \quad (84)$$

Ecosystem total vitality:

$$U(\alpha) = \bar{b}(\alpha)\bar{A}^a(\alpha)\Psi(\alpha) \quad (85)$$

Total vitality is a basic ecosystem goal function. It shows that ecosystem growth and development maximizes the product of the average genome complexity, species life span, and ecosystem total action. Note:

$$\bar{U}(\alpha) = \frac{\sum_i^p V(\alpha)_i}{p} = \frac{\sum_i^p \int_0^{\alpha} v(t)dt}{p} = \int_0^{\alpha} \frac{\sum_0^p v(t)}{p}dt = \int_0^{\alpha} \bar{v}(t)dt = U(\alpha) \quad (86)$$

8.2—ECOLOGICAL LAW OF THERMODYNAMICS (ELT):

As mentioned above, the ecological law of thermodynamics correlates or asserts the complementarity of the theories that reflect the highest flux of useful energy during ecosystem growth and development and those which reflect the highest system's organization and function. Jorgensen (2012) showed that when the exergy captured (taken from Kay and Schneider [1992], expressed as a percentage of solar radiation) is plotted versus the exergy stored (unit J/m2 or J/m3), calculated from the characteristic compositions of the focal eight ecosystems, a Michaelis-Menten-like plot is realized.

Now we would like to suggest that this plot represents the mathematical formulation of the ecological law of thermodynamics. Therefore, if we let exergy = $E_x(t)$ and exergy storage = $E_s(t)$, we identify the plot by this equation:

$$E_x(t) = cE_s(t) \quad (87)$$

Therein, c is proportionality constant, and t is time. We claim that the ELT, equation (87), can be deduced by taking the time derivative of the maximum action principle, equation (80), which yields equation (81):

$$\therefore \dot{\Psi}(t) = \overline{K}\,\overline{v}(t) \quad (88)$$

We mean that equation (87) and (88) are analogous because $E_x(t)$, from equation (87), is in direct correspondence with $\dot{\Psi}(t)$, which is the total matter-energy metabolized by the system. On the other hand, $E_s(t) = B \times \beta$ is in direct correspondence with $\overline{v}(t)$ = total matter-energy metabolized x genome physical complexity. Note that B is body mass and β is beta value, a measure of information (Jorgensen 2012).

Fath et al. (2001) used the network perspective to codify and unify ten ecological extremal principles: maximum power, maximum storage, maximum empower and emergy, maximum ascendency, maximum dissipation, maximum cycling, maximum residence time, minimum specific dissipation, and minimum empower to exergy ratio. They showed that these seemingly disparate extrema are all mutually consistent, suggesting a common pattern for ecosystem development. We suggest that the ELT in its mathematical formulation is the bridge that links the pattern's two focal points: the principles that pertain to maximization of the total system throughflow (exergy) and those pertaining to maximization of total system exergy storage (system biomass, organization, and function). If we assume that the first types of principles are different manifestations of metabolism and the second types are different manifestations of bioinformation, we can conclude that using metabolism and bioinformation as reference state these principles are unified on the basis of the ELT. And since the ELT is derivable from the maximum action principle, we could also conclude that these different ecological goal functions are different expressions of the maximum action principle. It follows that maximization of total vitality is the resultant that reflects the effects of the different goal functions. The increase of total vitality leads to the increase of the system's total matter-energy metabolized, which includes increase of biomass; the increase of the system's developmental functional complexity, which includes its network; the increase of the system's genetic information content; and the increase of the system's life span. Thus the increase of total vitality accounts for all three growth forms: biomass, network, and information. Moreover, it also accounts for the fact that k-strategists usually have a longer life span.

8.3—ECOSYSTEM GROWTH AND DEVELOPMENT:

To account for ecosystem growth and development, Elsheikh (2010) proposes the following assumptions:

a)   Based on the maximum action principle, or ELT, ecosystem growth and development is a process of increase of total vitality.

b)   Ecosystem growth and development is subject to the conservation of total bioinformation.

The ecosystem total bioinformation conservation is given by:

$$T(t) = \bar{U}(t) + Z(N(t)) = \text{constant} \quad (89)$$

$\bar{U}(t)$ is the ecosystem maximum total vitality at time t during successive generations, and $Z(N(t))$ is the ecosystem total natality density at time t during successive generations. And $N(t) = \dfrac{dp/dt}{p}$ is where p is the number of species.

Based on these assumptions, ecosystem growth and development can be represented by the following logistic equation:

$$\therefore \quad \frac{dp}{dt} = p(c - \int_0^t \frac{\dot{\bar{U}}\,dx}{Z_N'(N(x))}) \quad (90)$$

Therein, c is integration constant.

$$\text{Let } \Delta = (c - \int_0^t \frac{\dot{\bar{U}}\,dx}{Z_N'(N(x))}) \quad (91)$$

In the initial ecosystem growth phase, there is no significant increase in the species' total vitalities: $\dot{\bar{U}} \approx 0 \Rightarrow \Delta \approx c$ It follows that it is a phase of exponential growth. In the development phase $\dot{\bar{U}}(t) > 0$ and $\Delta > 0$, at the same time Δ is decreasing, it follows that when the ecosystem is fully mature:

$$\Delta = 0 \Rightarrow p(t) = p_{max} = \text{constant} \quad (92)$$

It is appropriate to note that the plant biodiversity on Surtsey Island, when plotted versus the year, gave approximately a logistic curve (Jorgensen 2006).

Note that, $\dot{\bar{\nu}}(t) \geq 0$, can be used to assess ecosystem health, and $\bar{\nu}_s(\alpha)$ can compare the ecological complexity or biocomplexity of different ecosystems. $\bar{U}(\alpha)$, being the ecosystem goal function, can be used to determine to what extent a given ecosystem is more evolved than a second one (i.e., having greater biomass, greater biocomplexity, and a longer life span).

Our current theory has some merits with respect to maximum power (Odum 1983), maximum eco-exergy, and maximum ascendency. Though eco-exergy and ascendency grasp the essence of biological organization, which is a fusion of matter-energy and information, they are not derived or based on a general theory of biology. On the other hand, though energy flux (emergy), or energy rate density, has something to do with complexity, it does not accommodate biosystems as information processors. To overcome these limitations, the current theory regards ecosystem dynamics as the phenotypic expression of genome dynamics, which has the following advantages:

i.   It deals with the pool of the species themselves directly as given by equation (90) or, in general, as described by the conservation of total bioinformation.

ii.   The genome is a whole that is not reducible to the arithmetic sum of its genes, so we evade reductionism.

Having said this, knowing that there is no complete theory, the merits of the above-mentioned indicators are not undermined. They are applied for the assessment of ecosystem growth and development, ecosystem health, ecological flow networks, and so forth. Moreover, several goal functions have been proposed—for example, structurally dynamical models (Jorgensen 2006). These models account for changes in certain parameters due to the adaptation or shift in the species composition. The parameters change by introduction of eco-exergy as a goal function in the model.

CHAPTER 9

TOWARD A THEORETICAL BIOLOGY

What is theoretical biology about? It is about a mechanistic deductive theory that explains and unifies ontogenetic development and phylogenetic evolution, a theory that reveals the secret or dynamical essence of living systems. It is about a theory that removes present-day apparent disparity or antagonism between physics and biology and explains the origin and evolution of life in a naturalistic plausible manner. It is about a theory that extends Darwinism and accommodates human sciences in a rational nonantagonistic manner, providing a more dynamic conception of human nature. It is about a theory that provides theoretical basis for understanding and explaining the origin and dynamics of sociocultural evolution, including the origin of morality and religion. Thus theoretical biology should facilitate a new foundation of human knowledge (i.e., facilitate holistic science, which is the most fundamental and most general of all sciences).

9.1—UNIFIED THEORY OF LIFE

It is clear that the recognition of the nature of life necessitates broadening or extending the concepts of information, matter, and quantum field and action principle. Based on this broader conceptualization, life emerges as a new type of quantum field; it is a quantum information fractal field. The new quantum field is an outcome of the product of the collective dynamics of ordinary physical fields optimized by golden ratio based DNA fractal geometry. So the nature of life and its secret reside in the DNA or genome as a quantum information fractal field that generates, in addition to weak electromagnetic waves, bioinformation oscillations through successive generations. In this perspective, a QIFFT is born, which accounts for both ontogeny and phylogeny by the same set of self-organization laws, unifying different biological phenomena. In particular, both ontogeny and phylogeny are processes of generation and assembly of minor attractors. The theory is also capable of transcending the fundamental barriers that shield life recognition: reductionism, entropy principle, quantum decoherence, and phase space coordinates barriers. Moreover, it also explains life's hallmarks, borderline, and puzzles.

Reductionism Barrier:

Some of the reductionism antireductionism dilemma basic arguments are causal closure and Nagel's bridge law. Reductionists, based on the causal closure, which asserts that no physical event has a cause outside the physical domain, claim the reduction of biology to ordinary physics. On the other hand, antireductionists, based on Nagel's bridge law, which is supposed to unite the primary science (e.g., physics) with the secondary science (biology), claim that such a requirement cannot be met. They claim that biology is an autonomous subject matter. According to QIFFT, both arguments are not necessary (that is, it is not necessary to reduce biology to ordinary physics as well as it is not necessary to look for a bridge law to connect both disciplines). What is necessary, according to QIFFT, is to broaden the ontological foundation of contemporary physical theory by broadening the concept of matter.

Thus matter has complementary properties: matter waves at high-mass density, microscopically, which facilitates ordinary physics, and bioinformation oscillations at high-information density, which facilitates quantum information fractal biology. This resolution to the problem or dilemma provides honorable relief for both opponents. Each of them, at least partially, considers his or her position as being always correct. The reductionist who feels that his or her claim that life is reducible to physics is correct, because QIFFT is a physical theory; at the same time, the antireductionist who claims that life is not reducible to physics is also correct, whereby, we mean by physics in this context ordinary physics, or inanimate physics. Perhaps one may ask surprisingly by so doing whether we have a physical or biological theory. Is biology reduced to physics or is physics reduced to biology? These questions have now lost their significance, dichotomy, and paradoxical nature because it seems we have reached the limit where the domain of theoretical physics coincides with the domain of theoretical biology.

Physics-biology complementarity has been proposed by several authors, including Menas Kafatos and Robert Nadeau (1990), Mihai Draganescu (online source), and Michael Conrad (1988). However, the unrecognition of what physically distinguishes life from nonlife has obstructed the efforts of these scientists. For example Draganescu, in his article "Deep Reality, Conscious Universe and Complementarity," wherein he explores some ideas from the book of Kafatos and Nadeau (1990), says, "The universe, which is a quantum system, is not only a quantum system but something more. It has complementary properties. It is quantic and conscious." According to them, as he tries to explain, consciousness of the universe is related to the wholeness that exists under the quantum realm, and this consciousness is located in the whole of the universe, namely in its roots. That is in the deep reality.

These authors propose a reality deeper than the quantum realm, having consciousness nature and two subtle properties of information and energy in order to substantiate the generalized complementarity of physics-biology. Conrad (1988) also states that in the natural world exists a quantum mechanical/biological dualism, analogous to the wave-particle dualism found in the ordinary quantum domain. The methods of the quantum physicist and of the biological sciences are seen to be alternative approaches to the understanding of nature, involving two districts modes of description that can usefully supplement each other, and neither on its own contains the full story. Since they do not recognize that the proposed complementarity necessitates broadening the concept of matter (that is, finding a new property of matter), these theorists either restrict complementarity to its methodological aspect or assume a deeper level of reality than the quantum level to which both biology and quantum levels be reduced. So it is still a reductionism attitude.

Thermodynamics Barrier:

Nicolis and Prigogine (1977) approached the problem of self-organization from a thermodynamics point of view (i.e., with reference to concentration gradients and temperature). According to the second law of thermodynamics, an isolated system reaches in time the state of thermodynamic equilibrium that corresponds to maximum entropy. In other words, in an isolated system, all motion usually comes to a halt because of various kinds of friction, differences of electric or chemical potential equalized, and temperature becomes uniform by transfer processes. After a time, the whole system fades away into a dead inert lump of matter. This is why generally entropy increase is regarded as a law of increase of disorder or disorganization. It does not account for the apparent self-organization characteristic of biological systems. This necessitates extension of thermodynamic theory to account for open instead of closed and isolated systems.

Nicolis and Prigogine (1977) formulated an extended version of the second law, applicable to both closed and open systems. The total change of entropy in an open system during a time interval dt could be written as follows:

$$dS = d_e S + d_i S \quad (93)$$

The latter, $d_e S$, is the entropy flux due to exchanges (of energy or matter) with the environment; $d_i S$ is the entropy production due to irreversible processes inside the system, such as diffusion, heat conduction, and chemical reactions. The second law yields:

$$d_i S \geq 0 \quad (= 0 \text{ at equilibrium}) \quad (94)$$

For an isolated system $d_e S = 0$,

$$\therefore dS = d_i S \geq 0 \quad (95)$$

Thus open systems, like biological systems, differ from isolated ones because of the presence of flow terms—d_eS in the entropy change. Contrary to d_iS, which can never be negative, these terms do not have a definite sign. The total change of entropy in an open system can be negative as well as positive. Therefore, we can imagine evolution, where the system attains a state of lower entropy than the initial one. This state, which from the point of view of the second law is highly improbable, can be maintained indefinitely, provided that the system is allowed to attain a steady state such as dS = 0 or the following:

$$d_eS = -d_iS < 0 \ (96)$$

Hence if we gave a system a sufficient amount of negative entropy flow, it would be able to maintain an ordered configuration.

Based on this vision, Kaila and Annila (2008) describe evolution as an energy transfer process, and since physical motion always takes the path of least resistance (i.e., the principle of least action), organisms can be depicted mathematically as dissipative systems that maximize the rate of entropy production in a system. Although an open system's energy landscape is in constant flux, they emphasize, that it always follows the most direct route (shortest path and steepest descent) to maximize rates of energy dispersal and entropy. Therefore, natural selection favors genetic mutations that lead to faster rates of entropy and consequently faster rates of evolution.

The importance of this view is that it emphasizes that evolution does not violate the second law of thermodynamics; it also utilizes the least action principle to conclude that evolution has traversed the shortest path. However, we must distinguish between thermodynamic order as a measure of physical complexity, whether in terms of bits or in joules/ degree kelvin, and biocomplexity or bioinformation that is developmental functional complexity measured in joules x bits. This distinction clarifies that evolution, if as we assume involves increase of bioinformation or biocomplexity, is only partially accounted for by the second law of thermodynamics, as it is obvious from the dimensions. This means the second law does not prohibit evolution as in increase of biocomplexity, yet it is insufficient to account for such an increase. Something more is needed (i.e., biological evolution), as we have tried to demonstrate is subject to the life-organizing principle, where it is driven by the maximum action principle rather than the least-action principle. In the next section, we prove that the least action principle is a special case for the maximum action principle.

Quantum Decoherence Barrier:

As it becomes clearer that the specificity of a complex biological activity does not arise from the specificity of the individual molecules that are involved (Van Regenmortel 2004). In fact, here come the important developments in coherence quantum electrodynamics from Del Guidice (1993) and Preparata (1995), who contend that condensed matter and living matter cannot be reduced only to their molecular components, but they can be reduced to the molecules oscillating in tune with an electromagnetic (EM) field. An example to show the interaction of an EM field in vacuum and matter is the Lamb shift, according to which the energy of an electron surrounding the proton in a hydrogen atom is slightly lower than the value calculated from the atomic theory based on purely static forces. Although this shift is very small, it provided evidence of the quantum vacuum fluctuation that has to be understood within the framework of quantum electrodynamics.

For a collection of particles, the usual approach is to apply the Lamb shift to each particle separately. While this is correct for low-density systems like gases, where the distance between any two particles is larger than the wavelength of the relevant fluctuating fields coupled to the systems, dense systems—condensed matter or liquids and solids—show entirely different behavior. When energy is absorbed from the vacuum field, the particles will begin to oscillate between two configurations. In particular, all particles coupled to the same wavelength of the fluctuations will oscillate in phase with the EM field—that is, they will be coherent with the EM field. Consequently, a coherent oscillation of matter and EM fields arises, described by a unique wave function in which it is impossible to trace the individual components. The wave function Ψ is characterized by the amplitude Ψ_0, whose square is proportional to the number of quanta of the field, and by the phase ϕ, which defines the rhythm of the oscillation of the field

$$\Psi = \Psi_0 \, e^{i\phi} \quad (97)$$

The state where the field acquires a defined value of ϕ is coherent.

Equation (97) defines a coherence domain for a population of atoms, molecules, and other microstructures.

Emilio Del Giudice and his colleagues (2011) emphasize that quantum fluctuations and coupling between matter and electromagnetic fields predict quantum coherence for liquid water even under ordinary temperatures and pressures. The theory suggests that interaction between the vacuum electromagnetic field and liquid water induces the formation of large, stable coherent domains (CDs) of about 100 nm in diameter at ambient conditions, and these CDs may be responsible for all the special properties of water, including life itself.

Referring to coherence quantum electrodynamics (CQEM) with regard to a decisive move toward coherence quantum biology, Ho (1995) says, "The explanations are tentative and incomplete in many respects. How do the original DNA molecule and their counter-ions interact with the water CD? Studies on DNA and protein hydration have revealed dynamic coherence between hydration water and macromolecule; although it is far from clear whether studies on macromolecules in solution can tell us anything about the macromolecules inside the living cell. Nevertheless, the quantum electrodynamics theory of water provides a useful framework for further investigations that decisively moves biology away from classical towards quantum physics and a much better understanding of non-thermal EMF effects."

Although neither classical nor standard quantum theory predicts quantum coherence for water, largely because they ignore quantum fluctuations and the interaction between matter and the electromagnetic field, which are taken into account in a quantum electrodynamics (QED) field theory (Ho 1995). It is also clear that QED field theory in turn ignores DNA and protein coherence generated by golden ratio based dodeca and icosa fractal geometry. The structuring and coherence of water that assemble a whole DNA sequence from its basic ingredients, as in the case of Montagnier's discovery, is accounted for by quantum information fractal field conjugation rather than by CQED, as mentioned earlier with regard to Montagnier's discovery.

Phase Space Coordinates Barrier:

Stuart Kauffman (1995), in his "At Home in the Universe; In Search for the Laws of Self-Organization," was hopeful with regard to discovering the laws of self-organization when he wrote, "In considering whether there can be laws of life, many biologists would answer with a firm no. Darwin has properly taught us of descent with modification. Modern biology sees itself as a deeply historical science. Shared features among organisms—the famous genetic code, the spinal column of the vertebrates—are seen not as expressions of underlying law, but as contingent useful accidents passed down through progeny useful widgets, found and frozen thereafter into the descendent branch of life. It is by no means obvious that biology will yield laws beyond descent with modification. But I believe that such laws can be found."

Kauffman suggests that when the complexity of the system (as represented by buttons and strings) reaches a critical threshold, new modes of organization can arise in the system "for free,"—that is, naturally and spontaneously. Kauffman illustrates that for a dynamical system, such as an autocatalytic net, to be orderly, it must exhibit homeostasis (i.e., it must be an attractor (1995, 79).

Kauffman summarizes the basic two features of his theory of Boolean network, which under certain conditions facilitates order for free. One feature is simply how many "inputs" control any lightbulb. If each lightbulb is controlled by one or two other lightbulbs, if the network is "sparsely connected," then the system exhibits stunning order. If many other bulbs control each bulb, then the network is chaotic. So "tuning" the connectivity of a network tunes whether one finds order or chaos. The second feature that controls the emergence of order or chaos is simple biases in the control rule themselves. Some control rules; the *and* and *or* Boolean functions tend to create orderly dynamics. Other control rules create chaos.

As we said earlier, this model suffers some inadequacies and is criticized by some authors. For example, Meyer (2004) indicates that in the light system, the order that allegedly arises for "for free" actually arises only if the programmer of the model system "tunes" it in such a way as to keep it from either (a) generating an excessively rigid order or (b) developing into chaos. Meyer emphasizes that this necessary tuning involves an intelligent programmer selecting certain parameters and excluding others—that is, inputting information. In addition, Kauffman's model systems are not constrained by functional considerations and thus are not analogous to biological systems. Moreover, based on Elsheikh's (2010) proposed generalized complementarity, a material system, animate or inanimate, does not possess simultaneously both matter waves and bioinformation oscillations descriptions. This means that models based on inanimate processes like the sand pile and lightbulbs do not contain the dynamical essence of living systems.

Thus in a paper titled "No Entailing Laws, but Enablement in the Evolution of the Biosphere," Longo, Montévil, and Kauffman (2012) express a new different vision. They wrote, "The aim of this article is to demonstrate that the mode of understanding in physics since Newton, namely differential equations, initial and boundary conditions, and then integration, which constitutes deduction, which in turn constitutes 'entailment', fails fundamentally for the evolution of life. No law in the physical sense, we will argue, entails the evolution of life."

They claim, if we are correct, that this spells the end of "strong reductionism," the long-held belief that a set of laws "down there" entails all that happens in the universe. Moreover, if no law entails the evolution of life yet the biosphere is the most complex system we know of in the universe, it has managed to come into existence without an entailing law. Then such law is not necessary for extraordinary complexity to arise and thrive. Therefore, they emphasize the need for new ways to think about how current life organization have come into being and persisted.

The heart of their considerations:

1) In physics, the phase space that contains the system's dynamics can be prestated.

2) In biological evolution, the phase space itself changes persistently. More, it does so in ways that cannot be prestated.

3) Because we cannot prestate the ever-changing phase space of biological evolution, we have no settled relations by which we can write down the "equations of motion" of the ever-new biologically "relevant observables and parameters" revealed after the fact by selection acting on Kantian wholes in biological evolution, but that we cannot prestate. More, we cannot prestate the adaptive "niche" as a boundary condition, so we could not integrate the equations of motion even were we to have them.

4) If the above is true, no law entails the evolution of the biosphere.

5) If by "cause," we mean what gives a differential effect *entailed by law*, then we can assign no cause in the "diachronic" evolution of the biosphere.

6) In place of "cause" in this diachronic evolution, we will find "enablement" (i.e., making possible—a key notion in our analysis).

7) Our thesis does not obviate reductive explanations of organisms as synchronic entities, such as an ultimate

physical account of the behavior of an existing heart, once evolved.

How does QIFFT overcome these difficulties? First, we celebrate with Kauffman and his colleagues the end of strong reductionism, if by physics we mean ordinary physics. Second, we agree that the phase space coordinates do not contain the dynamical essence of living systems. The breakthrough is to realize that both ordinary physics and its phase space coordinates do not contain the whole domain of physical reality. So Kauffman proposed that organisms as Kantian wholes could only be represented within a holistic domain of physical reality. This necessitates a new quantum information fractal field that contains the dynamical essence of living systems other than ordinary physical phase space. Within such broadening of the ontological foundation of contemporary physics, Kauffman's (1995) initial conviction on laws of self-organization can be regained and realized. Because what generates the dynamics of a living system is its internal quantum information fractal field, a function over bioinformation and time, rather than the external physical fields, we do not need phase space coordinates in the ordinary sense, i.e. the system's bioinformation (vitality) is independent of space coordinates.

Explanation of Life's Hallmarks:

To compare and evaluate accounts of the nature of life—according to Bedau (2003)—necessitates explaining life's hallmarks, explaining the borderline cases, and resolving the puzzles about life. In other words, a major task of a theoretical biology is to explain these three things. It is clear that such an explanation depends on how life is defined. Whereby, we consider a system alive if it possesses a quantum information fractal field that generates bioinformation oscillations through successive generations, subject to the first and second laws of self-organization. Moreover, a living system is said to be alive if it satisfies the conditions (46):

$$\dot{\Phi}(0) > 0 \ \ and \ \ v(0) > 0$$

On these bases, Ganti's real and potential hallmarks or criteria can be explained. These are:

i-   Real (absolute) life criteria

a. A living system must inherently be an individual unit. (Our analysis is based on the organism as a whole.)

b. A living system has to perform metabolism. (The first law of self-organization identifies the metabolic nature of the organism and its relatedness to the system's bioinformation.)

c. A living system must be inherently stable, despite changes in the external environment. (The life-organizing principle is a structurally stable system.)

d. A living system must have a subsystem carrying information that is useful for the whole system.

(We demonstrate that the genome is a quantum information fractal field.)

ii- Potential life criteria

a. A living system must be capable of growth and reproduction. (We demonstrate that the system exhibits self-sustained bioinformation oscillations.)

b. A living system must have the capacity for hereditary change and, furthermore, for evolution. (The life-organizing principle and the second law of self-organization have the capacity for heredity change and predict the new evolved quantum stationary functional state.)

c. Living systems must be mortal, and nonliving systems cannot die, so death is characteristic of life. (Life-organizing principle boundary conditions indicate that a living system has no life before birth or after death.)

Familiar examples for Bedau's borderline hallmarks are viruses, which self-replicate and spread even though they have no independent metabolism. Moreover, there is the case of dormant seeds or spores and even frozen bacteria or insects.

As we said, viruses are living because they satisfy the conditions (46):

$$\dot{\Phi}(0) > 0 \ \ and \ \ v(0) > 0$$

However, this is what is special about viruses:

$$\dot{\Phi}(0) = v(0) = cons \tan t > 0$$

So $\ddot{\Phi}(0) = \dot{v}(0) = 0$, which means the system does not metabolize.

It is also clear from (49) that viruses or RNA replicators, in general, possess weak stability (i.e., the stability of a conservative system, which can easily be changed by positive damping or negative damping. This may explain the ease with which a virus can change its biological properties.

Some scientists do not classify viruses as living even though they can replicate and evolve, because they do so only within cells and lack important cellular attributes, such as metabolism and irritability. According to the present theory, matter has two classifications: inanimate, which do not generate bioinformation oscillations, and replicators, which generate bioinformation oscillations. Although replicators could be classified in two phases—replicators that generate autonomous self-sustained bioinformation oscillations and replicators that do not generate autonomous self-sustained bioinformation oscillations—they do replicate and evolve, particularly RNA replicators. Nonetheless, as far as these replicators (e.g., viruses) satisfy conditions (46), they are considered alive.

The case of dormant seeds or frozen insects can be explained by indicating that $\dot{\Phi}(0)$ *and* $v(0)$ are not always coupled; there are cases or systems for which $v(0) = 0$ but $\dot{\Phi}(0) > 0$ such systems are dead though their action increases (e.g., an accelerating car). However, there is the case in which v(0) > 0, but $\dot{\Phi}(0) = 0$; this is the case with dormant seeds or frozen insects. Such systems regain their vitality increase whenever favorable environmental conditions are realized.

With regard to what Bedau called life puzzles, we propose that the origin and emergence of life are to be considered in view of QED and autocatalysis. The QED facilitates fractal field phase conjugation and coherence based on the RNA dodeca fractal geometry, whereas autocatalysis facilitates the kinetic power of replication. Other puzzles like the hierarchical and discreteness of living systems are accounted for by revealing the fractal and quantum nature of living systems. Artificial life systems (soft or hard), according to the present theory, are not living systems because they do not satisfy the proposed life criterion, i.e., possessing quantum information fractal fields that generate bioinformation oscillations for successive generations, subject to the first and second laws of self-organization. With regard to the mind as a puzzle, we consider the mind intelligence, and intelligence is bioinformation, so mind's intelligence is life's or nature's intelligence.

9.2—LIMITING TRANSITION TO QUANTUM MECHANICS

Antoniou and Prigogine (1993) differentiate between extrinsic and intrinsic irreversibility. A system possesses extrinsic irreversibility as a result of being coupled to environment. If decoupled, it then becomes reversible. On the other hand, intrinsic irreversibility is self-generating by the internal dynamics of the system. According to Antoniou and Prigogine, intrinsic irreversibility is a fundamental attribute of certain dynamical systems that generate order away from equilibrium.

Now, although biosystems possess intrinsic irreversibility since biotic growth and development drive the system away from thermodynamic equilibrium, intrinsic irreversibility in this sense is not sufficient to account for biotic irreversibility. An important attribute of biosystems is that they evolve, which means they maximize total vitality; therefore, they maximize the involvement of time in the life process. It follows that biotic irreversibility is not only self-generating but also self-evolving (that is, it goes on driving the system away from thermodynamic equilibrium).

Thus for the same time arrow, there are two different types of irreversibility: maximization of entropy, which drives the system toward thermodynamic equilibrium, and maximization of total vitality, which drives the system away from thermodynamic equilibrium. Evidently, creative processes (bios) are fundamental in nature as well as destructive processes (Thomas et al. 2006). The difference between systems that possess mere intrinsic irreversibility and biotic systems is that the former do not evolve. They are incapable by themselves of continuing to navigate away from thermodynamic equilibrium, the navigation or evolution that maximizes the involvement of time in the life process.

Basic Findings of QIFFT:

(a)— $L = L_0 e^{\frac{i}{k}\int v(t)dt}$, where k = bGA^a, G is a constant.

And $v(t) = bE(t)(A-t)^a = bE(t)\,\ell^a$, t $\geq$ 0

(b)—V(A) = $\int_0^A v(t)dt$ = constant,

Under constant environmental conditions.

Proof: $V(t) = \int_0^t v(x)dx$

Given:

$\therefore \dot{V}(t) = v(t)$

$\therefore \dot{V}(A) = v(A) = 0$

$\therefore V(A)$ = constant, (98) under constant environmental conditions.

The life-organizing principle being structurally stable, the topological character of the vitality curve is preserved through successive generations. Hence total vitality is conserved. The conservation of total vitality indicates that the DNA or genome is in a stationary state, so the species reproduces its own kind.

(c)—V(A) = $\dfrac{n\pi \overline{G}}{f^a}$ n = 1, 2 . . . , and f is the frequency of bioinformation oscillations.

(d)—$\Phi(t) = K\int_0^t v(x)dx$

Limiting Transition

To derive the basic laws of linear reversible quantum mechanics as a special case from QIFFT, we begin by introducing the following assumptions:

Assumption (A):

Biotic irreversibility is the fundamental attribute for the transition from inanimate to animate systems.

Assumption (B):

The time t—the age of organism—measured from the moment of initial growth is at the same time a measure of biotic irreversibility.

Note that in this case, the time t has two meanings. It is a measure of the organism's age as well as a measure of the organism's bioirreversibility. For example, when the organism dies at age t= A, its bioirreversibility is t = 0.

Since the time t is a measure of bioirreversibility, then the main idea is to eliminate the bioirreversibility from the equations of QIFFT by taking the limit as t goes to zero in order to obtain the corresponding inanimate laws belonging to the lower level of the physical hierarchy. For this purpose, let the vitality of a dead system be given by v(0), which means the system's bioirreversibility is t = 0. We assume for such a dead or inanimate system that its total matter-energy content E(0) is constant, i.e., subject to the law of conservation of matter-energy.

Theorem:

The laws of QIFFT admit limiting transition to linear reversible quantum mechanics.

Proof:

Let t $\rightarrow$ 0 $\Rightarrow$ L$\rightarrow$ Ψ and G$\rightarrow$ $\hbar$, under limiting transition, then we get:

$$L = L_0 e^{\frac{i}{k}\int v(t)dt}$$

$$\therefore \underset{t\to o}{Limit}\frac{d}{dt}(ikL) = \underset{t\to 0}{Limit}\frac{d}{dt}\left[ikL_0 e^{\frac{i}{k}\int v(t)dt}\right]$$

$$\therefore \quad i\hbar b A^a \frac{d\Psi}{dt} = -v(o)\Psi = -bE(0)A^a\Psi$$

$$\therefore i\hbar\frac{d\Psi}{dt} = -E(0)\Psi = -\overline{\overline{H}}\Psi \quad (99)$$

Having been able to specify the system's Hamiltonian, $\overline{\overline{H}}$, one may identify the corresponding Schrödinger's equation, whose general solution is:

$$\Psi(r,t) = \Phi(r)e^{\frac{i}{h}\int E(0)dt} \quad (100)$$

(b) $V(A) = \int\limits_0^A v(t)dt = cons\tan t$

$$\therefore \underset{t\to 0}{Limit}\frac{d}{dt}V(A) = \underset{t\to 0}{Limit}\frac{d}{dt}\int\limits_0^A v(t)dt = \underset{t\to 0}{Limit}\frac{d}{dt}const. = 0$$

$$\therefore \underset{t \to 0}{Limit} \int_0^A dv \;=\; 0 \;\Rightarrow\; \underset{t \to 0}{Limit} \left[v(A) - v(t) \right] \;=\; const.$$

$$\therefore \; -E(0)bA^a \;=\; const. \;\Rightarrow\; E(0) \;=\; const. (101), \text{ under}$$

constant environmental conditions.

Thus conservation of total vitality yields in the limiting case the law of conservation of matter-energy.

(c) For simplicity, if we assume that the exponent a is approximately constant along a given phylogenetic lineage, we get:

$$V[A(t)] \;=\; \int_0^{A(t)} v(x,t)dx \;=\; \frac{n\pi b\,G}{f^a} \;=\; n\pi \, bGA^a(t)$$

$$\therefore \; \underset{t \to 0}{Limit} \frac{dV}{dt} \;=\; \underset{t \to 0}{Limit} \left[\int_0^{A(t)} dv(x,t)\frac{dx}{dt} + v(A,t)\frac{dA}{dt} \right] \;=\; n\pi \, bG \underset{t \to 0}{Limit} \frac{d}{dt} A^a(t)$$

we let $\dfrac{dA(t)}{dt} = const. = 1$, where A(t) $\propto$ t. Also note that b is

basically inanimate property, measured in bits.

$$\therefore \; \underset{t \to 0}{Limit}[bE(t)(A(t)-t)^a - bE(t)(0-t)^a + bE(t)(A(t)-t)^a] \;=\; n\pi \, b\,Ga \underset{t \to 0}{Limit} A^{a-1}(t)$$

$$\therefore \, 2E(0)A^a(0) \;=\; n\pi \, GaA^{a-1}(0)$$

$$\therefore \; E(0) \;=\; \frac{n\pi \, a\,G}{2A(0)} \;=\; \frac{1}{2} n\pi \, aGf(0)$$

If $t \to 0 \;\Rightarrow\; \dfrac{1}{2}\pi aG \to h,$ then we get Planck's law:

$$E(0) \;=\; nhf(0) \quad (102)$$

It follows that quantization of total vitality yields in the limiting case energy quantization.

$$(d)—\Phi(t) = K\int_0^t v(x)dx$$

The stationary action principle is a special case of the principle of maximum action.

Proof:

A dead organism's total matter-energy content E(0) being subject to law of conservation of energy, we let:

$\therefore$ E(0) = constant

From (d) follows,

$$\therefore \dot{\Phi}(t) = K v(t)$$

$$\therefore \lim_{t\to 0}\dot{\Phi}(t) = \lim_{t\to 0}Kv(t) = KbE(0)A^a = \frac{bA^a E(0)}{bA^a} = E(0) \quad (103)$$

$$\text{Let } \lim_{t\to 0}\dot{\Phi}(t) = \frac{dS}{dt}$$

$$\therefore \frac{dS}{dt} = E(0) \Rightarrow S = \int E(0)dt \quad (104)$$

Equation (104) defines the principle of least or stationary action for a conservative system.

It is interesting that the proposed QIFFT, in the limiting case, leaves the basic laws of linear reversible quantum physics invariant, which supports Elsasser's (1988) view when he says, "There is no shred of evidence anywhere in the vast literature of biochemistry or biophysics that the laws of physics (in practice the laws of quantum mechanics) are invalid or stand in need of modification. Any approach to theoretical biology must start from this basic fact. But at the same time this does not absolve us from the need to pursue the possibility of a substantial conceptual innovation occurring in the passage from theoretical physics to theoretical biology. In other language, any conceptual innovation must be such that it leaves laws of quantum mechanics invariant." It seems that by saying "in the passage from theoretical physics to theoretical biology . . . any conceptual innovation must be such that it leaves laws of quantum mechanics invariant," Elssaser envisages the possibility of a QIFFT.

9.3—NEW FOUNDATION OF HUMAN KNOWLEDGE

Present-day human knowledge is in disarray, perplexed by dichotomies and anomalies: reductionism and antireductionism, materialism and idealism, nature and nurture, structure and history, and so forth. The grassroots for all mentioned dichotomies is that a fundamental property of matter escaped human imagination and investigation for centuries. The discovered new property (i.e., bioinformation oscillations) broadens the ontological foundation of contemporary physical theory and reveals the hidden intrinsic complementarity nature of the universe. Based on the proposed generalized complementarity of matter waves and bioinformation oscillations, matter and phenomena are characterized by the unity and complement of opposites.

This is different from materialistic monism, which reduces mind to body, or monistic idealism, which does the opposite.

It is different from Cartesian dualism, which asserts the independence of these opposites, disregarding their actual complement. It is also different from dialectical materialism, which claims the unity and struggle of opposites. This is because dialectical materialism is a monistic philosophy for which the identity principle prohibits the complement of opposites. Based on the identity principle, a proposition is either true or false; it cannot be both. Complementarity does not exclude the possibility that the two mutually exclusive statements could identify the proposition, but not simultaneously. For example, an electron is not either wave or particle; it is both, but not simultaneously. Now, according to QIFFT, matter has two complementary properties: matter waves and bioinformation oscillations. So the oneness of the universe is intrinsically embedded in its complementarity, the rich source of its unlimited diversity and differentness. Thus complementarity fosters diversity and differentness in contradistinction to totalitarian excludable discourses and ideologies.

The proposed generalized complementarity resolves reductionism-antireductionism dichotomy in a peaceful manner. Each opponent can now claim victory, or at least each of them feels that his or her position is partially correct.

The solution to the contradiction between biology and physics also contributes to solving the contradiction between biology and human sciences. "Thus scholars in the humanities and social sciences have long been trained to believe that biology and human biology are practically irrelevant to their subjects" (Azar Gat 2006). On the other hand, sociobiology, psychobiology, and evolutionary psychology base their view of human nature on Darwinian instinct. According to Darwin's theory of natural selection as an account of how evolution takes place, it is not difficult to explain why there is selfishness, violence, and despotism. These can be taken for granted as a measure of reproductive fitness. The problems become how to explain morality and altruism. Here comes the important discovery of QIFFT—that the genome total bioinformation generates two survival components: reproductive fitness component and total vitality fitness component. This extension of Darwinian theory by accommodating a new fitness component substantiates the theory of multilevel selection by identifying the group selection fitness unit. It also facilitates the discovery of the complementary nature of human nature.

According to the QIFFT, phylogenetic evolution maximizes total vitality. By so doing, based on the conservation of genome total bioinformation, it decreases reproductive fitness. If we associate the decrease of reproductive fitness with increase of altruism, we conclude that phylogenetic evolution maximizes altruism. This is quite evident from the evolution of major transitions. Wilson and Wilson (2006) emphasize that there is agreement that selection occurs within and among groups, that the balance between levels of selection can itself evolve, and that a major transition occurs when selection within groups is suppressed, enabling selection among groups to dominate the final vector of evolutionary change. The major transition in evolution refers to the transitions from solitary replicators to a network of replicators enclosed within compartments, from independent genes to chromosomes, from prokaryotic cells to eukaryotic cells containing organelles, from unicellular to multicellular organisms, and from solitary organisms to colonies (Okasha 2005). On the other hand, since increase of total vitality is associated with bioinformation, which is creativity, the ability of the organism to adapt or change environmental conditions to suit its well-being, we conclude that phylogenetic evolution maximizes both creativity and altruism. We call creativity and altruism faeeliya, which is the group selection fitness unit. Biological evolution as well as human sociocultural development (evolution) maximizes faeeliya (i.e., maximizes creativity and altruism) (Elsheikh 2005).

We assume that due to the evolution of Homo sapiens brain architecture and the emergence of human higher mental powers, faeeliya has been realized as an ontogenetic property. Thus humans become both products and producers of faeeliya. This reveals the complementary aspects of human nature being subject to reproductive fitness maximization as well as to faeeliya maximization (this explains humans' complementary aspects of evil and virtue). Humans, being producers of faeeliya, signifies and originates the sociocultural evolution and at the same time explains why human sociocultural evolution maximizes faeeliya faster than genetic evolution. Accordingly, Wilson and Wilson (2006) state the following summary of sociobiology's new theoretical foundation: "Selfishness beats altruism within groups. Altruistic groups beat selfish groups. Everything else is commentary."

A human can be at low faeeliya or high faeeliya. At low faeeliya, he or she expresses the Darwinian instincts and attributes of reproductive fitness (i.e., selfishness, violence, and despotism). At high faeeliya, he or she expresses the faeeliya fitness attributes of creativity and altruism. Of course sometimes, due to the influence of social pressure and moral obligations on mind structures' dynamics, a human may fluctuate between high and low faeeliya. Then the following important questions arise: What do we mean by low and high faeeliya? What is the mechanism for faeeliya development or increase? To answer these questions, we need to propose a theory of mind structures dynamics.

Mind Superimposed Structures:

It is appropriate to assume that in human perspective, faeeliya is manifested through the expressions of differentiated mind structures superimposed upon one another to support the basic human life drives and needs, which are:

i.   Reproduction

ii.   Production and possession of means of subsistence

iii.   Holistic protection and enrichment of life

Thus we get the following mind structures:

Reproduction Mind Structure (RMS):

The RMS secures human life by giving priority to reproduction. Consequently, a person conceives of himself or herself as a reproductive being whose main role in life is to give birth and to care for as many children as possible.

* Materialistic (Bourgeois) Mind Structure (MMS):

This mind structure gives priority to producing and possessing means of subsistence. Thus a person conceives himself or herself as a materialistic or economic being.

* Creative Mind Structure (CMS):

Through this mind structure, a person conceives of himself or herself as a humane and creative being whose role in life is characterized by love, creativity, and altruism. In general, mind structure is consciousness-generating structure, which determines one's self concept and one's attitude and response toward sociocultural challenges.

Every individual in every society possesses all three of these mind structures. However, the dominance of a given mind structure of any social group is determined by the mind structure's ability to respond to the most threatening socio-historic challenges. The dominant mind structure usually utilizes the other two mind structures to achieve its own goals and means. Therefore, the mind structures cooperate when one of them produces a positive response to the concurrent problems of social development, security, and survival. However, if the dominant mind structure fails to respond positively and effectively, the remaining two mind structures will compete for the opportunity to give a better response. The RMS dominates when there is no means to combat high death rates other than maximizing birth rates and when man's sheer physical power is the source of social and economic security.

The MMS starts to dominate when RMS accomplishes its mission and cities become populated to the extent that acquiring means of subsistence becomes a major challenge—and when MMS and CMS cooperate to overcome the challenge. It is important to note that both RMS and MMS have lower or partial faeeliya, because their programs for love are closed (i.e., love is confined to oneself and/or to close relatives). The CMS is the only open mind structure that provides love and support to all people without discrimination. The CMS therefore possesses high faeeliya and can actualize the universal values: freedom, justice, equality, and altruism. Each mind structure has its own referential system of values, referential system of knowledge, and referential self-concept that facilitates the accomplishment of its project. It is also important to emphasize that the basic attribute of mind structures is their dynamics and interchangeability, on an individual level as well as a social level. So the dominance of a given mind structure is by no means complete or immutable.

Faeeliya Analysis:

It is a method by which the faeeliya of individuals, societies, and literary texts is revealed—revealing mind structures' dynamics (development and interaction) as they respond to socio-historic challenges.

Faeeliya Development Mechanism:

Faeeliya development means the transcendence (not abolition) of RMS and MMS objectives and regimes (projects) and acquiring CMS project of creativity and altruism according to the following mechanism:

A-   Challenge: one may suffer from different obstacles and challenges throughout one's life, in particular focusing on psycho-existential insecurities, for psycho-existential sufferings and insecurities are the first step to being in disharmony with the prevailing mind structure and its dominant discourse and project.

B- Breakdown of the societal dominant mind structure (i.e., fails to provide basic human life securities).

C- Response, which means breaking away from the dominant societal mind structure—which is either RMS or MMS—with a faeeliya project and fostering the project. There is no faeeliya development without faeeliya project development. Simultaneously, the process of faeeliya development is a process of faeeliya universal consciousness development. There is no faeeliya project development without perseverance, no perseverance without power of hope, and no power of hope without faeeliya universal consciousness. faeeliya consciousness is life's ultimate actualization and empowerment; it is life being conscious of her universal efficacy, creativeness, and holistic enrichment. Suffice to say, there is no creativity without faeeliya development. Creativity is not mere intelligence; it is essentially faeeliya: the discovery of a universal order, the signature of ultimate altruism, necessitates a universal ego, vibrant with universal love. (Elsheikh 2005, 2013)

This mechanism is applicable to individuals as well as societies. For example, social structure is a superimposition of mind structures under the dominance of one of them, which is relative dominance. Therefore, we get RMS, MMS, or CMS social structures. Social change is, then, a transformation from one dominant mind structure to another, in accordance with faeeliya development mechanism. The transformation takes place when the new social forces that incorporate the new mind structure become a major force. In this manner, social structure is dynamized. Thus a theory of historic development must be based on the dynamics of social structures.

Because altruistic groups beat selfish groups, the social need for faeeliya development and faeeliya fitness creates both morality and religion. This is why the basic doctrine or command for almost all religions is "love thy neighbor." However, for any religion to survive, it must be multilevel discourse, which addresses all mind structures. So each person interprets it and interiorizes it according to his or her own dominant mind structure and to his or her own level of faeeliya. It follows that religion is a unifier and a potential subdivider. Religious violence and other-disregarding behavior is an expression of low faeeliya. This is why Christianity started as pure CMS discourse, however, later on incorporated low faeeliya discourse to match the other mind structures. "Christianity, starting as a religion of love, compassion, and nonviolence, later developed a brutal militant streak toward non-believers and heretics, which awkwardly but continuously coexisted with its opposite in both doctrine and practice" (Azar Gat 2006). faeeliya vision, by extending Darwinism and revealing the complementary nature of human nature, reconciles the existing contradiction between science and the final goals of religion.

The Darwinian Phase

Biological evolution as well as sociocultural evolution, being maximization of faeeliya, indicates that human sociocultural evolution is a transition from stages of low faeeliya to stages of high faeeliya. Thus the early stages of this historic movement would be dominated by low faeeliya characteristics: selfishness, violence, and despotism. One finds no difficulty in finding evidence for the brutality and savageness of human nature since people were hunter-gatherers up to now. Despite First and Second World Wars, wars are everywhere. Look what happened in Kosovo, Afghanistan, Rwanda, Iraq, Dar Fur, Syria, and so on. In addition to wars, domestic violence prevails everywhere: violence against women, violence against children, violence due to racism, violence due to exploitation, violence due to marginalization and exclusion. Moreover, violence against human species and life as a whole is mounting by destabilizing and destroying the biosphere. For this reason, we call this phase of human sociocultural evolution the Darwinian phase. We call the high faeeliya phase that transcends the Darwinian phase just the faeeliya phase; it is the phase in which the CMS is dominant.

During the Darwinian phase, the dominant societal structures are RMS first and then MMS, which characterizes the rise of contemporary Western civilization. However, the majority of the Third World countries are still dwelling under the influence of RMS dominance, but some of them are undergoing a transitional phase toward MMS. Although MMS societies represent higher faeeliya than RMS societies—due to higher standards of living, human rights concern, democracy concern, social care, technological advancement, and so on—these societies are still dwelling in the Darwinian phase because in these societies less than 20 percent of the population monopolizes more than 80 percent of the societal wealth and power, so there is no just distribution of wealth and power. Moreover, although these countries departed the colonial era and started to show more concern for democracy and human rights as well as offering help during war crises, natural disasters, famine, and so on, to underdeveloped countries, they still keep unjust economic ties with these countries.

Although the suffering of the majority of population in a capitalist society is evident, with excessive inequality, unemployment, overwork, poverty, lack of democracy, and environmental degradation (Schweickart 2011), the people are not reluctant. Not only that, but capitalism actually proved to be more efficient economically than the Soviet style of socialism. The reason is that the majority of population in both systems belongs to the Darwinian phase, where private ownership and free market economy are basic attributes, according to which the individual works harder for his or her own benefit than for the sake of group or society. Human nature determines the nature of economy, not vice versa, as critical theory in general assumes.

Marxism claims that private ownership originates due to the development of productive forces, while private ownership was already there in primitive societies, where it took the form of possessing as many women as possible to give birth to men who could protect the family or tribe. Marxism disregards the biological aspect of human nature because according to mainstream Darwinism, the Darwinian instincts are the sole representative of human nature; such a claim justifies exploitation, as far as there is no faeeliya. However, by disregarding the biological aspect of human nature Marxism's vision of social change becomes inadequate, because the majority of proletariats may not have sufficient high faeeliya to create a socialist or postcapitalism society. It is also difficult for Marxism, without recognition of mind structures dynamics, to explain why some bourgeois like Engels, who supported and collaborated with Marx, have high faeeliya?

In the absence of high faeeliya in proletariats and peasants, the Communist party used violence to impose socialism in the Soviet Union, not only during the revolution but also after the revolution; violence itself is an attribute of the Darwinian phase. Such violence would only reproduce the Darwinian phase—it is Darwinian phase preserving transformation— and consequently fails to maintain socialism, for the simple reason that socialism or postcapitalism cannot be maintained within a Darwinian phase. The Darwinian phase is a favorable incubator of capitalism; in the majority of people, those who have not have the same nature and ambition as those who have. For example, pick at random a laborer and give him a sufficient amount of money. Would that person refuse to be a capitalist? Suppose he or she refuses how much percentage wise would refuse? This means that capitalism is ideal for a significant portion of laborers and the poor, which objectifies their attitudes and dreams, despite their sufferings or maybe because of their sufferings, and for this reason, the system survives, despite its flaws. This indicates that oppression is not a sufficient condition for the oppressed to have a different human nature than the oppressor, for they are both subject to the Darwinian phase.

Transcending capitalism necessitates transcending the Darwinian phase. But can humans transcend the Darwinian phase? The answer is yes, for human nature is complementary and dynamic. A postcapitalism society is by definition a CMS society, a high faeeliya society, and high faeeliya can be obtained by faeeliya development. It is also important to highlight that the dominance of RMS or MMS does not mean that the society does not possess high faeeliya representatives. They are the minority, but they are always there, due to faeeliya development mechanism. It is also important to note that although a low faeeliya person is programmed to manifest the dominant societal structure project and values, he or she possesses dormant high faeeliya, so in a sense you may regard a low faeeliya person a victim. For example, the oppressed laborer who dreams of being a capitalist is a victim, as is the capitalist. They are both victims of a certain sociocultural development phase. For a human society to transcend to a high faeeliya phase and maintain a postcapitalism society, the following conditions are necessary:

- **Faeeliya system of knowledge:** It is an open system of knowledge that does not designate nonbelievers or opponents as heretics or reactionaries; such a system is not monistic, because a monistic discourse or ideology is inevitably totalitarian and excludable. It is a faeeliya referential system of knowledge, which is a complementarity based system of knowledge. Another important issue is the reference of faeeliya (i.e., causal efficacy, whether intrinsic or extrinsic). The social actor must be a producer of faeeliya, not a beggar of faeeliya from historical determinism or whatever gods may be. This condition of intrinsic causal efficacy, which differentiates faeeliya consciousness from both materialistic and metaphysical forms of consciousness, is necessary for faeeliya development. It is necessary for the social actor to be aware of her or his activeness and faeeliya.

- **Faeeliya cultural project:** By definition, faeeliya is total enrichment and protection of life. This can be achieved by create universal love project (CULP) and create universal love movement (CULM). The CULP representatives are already there: human rights activists, labor movement activists, feminism movement activists, political activists fighting for democracy and social justice, peace movement activists, charity organizations, Doctors Without Borders, environmental protection activists, Red Cross activists, social welfare activists, and many others. These are in addition to individuals who participate in the welfare of humanity: scientists, artists, philosophers, writers, and so forth. When all these activists organize and cooperate locally and on a global scale, they then form what I call CULM. The movement advocates and supports all of the above-mentioned activists' causes. Its basic objectives are to develop methods and provide solutions that initiate and originate postcapitalism society in theory and practice, including development of faeeliya psychology, faeeliya sociology, and faeeliya economy. Within this project, the survival of the individual is not independent of the survival of the species and life as a whole. We are connected.

- **Faeeliya development:** There is no faeeliya development without faeeliya project development. So for a social actor, in addition to the person's work to improve his or her living standard, it is important to engage in a CULP, thereby working to improve the life of the whole. Why do I care for others? This question is a clear signature for low faeeliya, because at low faeeliya the ego is closed; it doesn't conceive its universality. So within the Darwinian phase, within low faeeliya, we either sacrifice the individual for the sake of society or sacrifice the society for the sake of individual.

This dilemma could be overcome by faeeliya consciousness development, whereby the individual unfolds her or his CMS universal ego of universal love, hence transcending the contradiction between the person's survival and group survival. The processes involves change of the referential system of knowledge, and change of the referential system of values, and change of self-concept—change from RMS or MMS self-concept to CMS self-concept. What one usually regards as one's self-concept or identity is actually a product of social programming i.e., product of a certain dominant societal structure.

In fact, due to mind structures dynamics and faeeliya development mechanism, there is no iron law that a capitalist must be selfish or completely governed by the laws of market economy. Under certain circumstances, a capitalist or a wealthy person may develop high faeeliya and use some of his or her wealth for humanitarian causes. Although this does not resolve the inadequacy of capitalism, it proves that CMS is there in every individual. If this is so, then it is probable that a day will come when some capitalists may contribute significantly to transcending capitalism. That day will demonstrate that faeeliya is the ultimate meaning and purpose of human life. That day will signify the glory of humanity. It is also remarkable that while a mind that can be programmed by misuse of its sociability and altruism to be a suicide bomber killing innocent social beings, the same mind, through faeeliya consciousness development, can participate in holistic protection and enrichment of life, which highlights the importance of developing faeeliya psychology.

New social force: It is evident that social forces of RMS origin or MMS origin cannot facilitate a transition to faeeliya phase or postcapitalism society; this explains why both Modernism and Marxism fail to actualize their dreams in a society of freedom, equality, and fraternity. From where do the new social forces arise? The new social forces arise due to a breakdown of the dominant societal structure, its failure to provide basic social, economic, and existential securities. In this regard, the new social forces are already there with respect to transcending capitalism. They are the CULP forces. What is needed is for these forces to organize and engage in faeeliya development on individual and social levels, also initiating or creating faeeliya space within the Darwinian space, bearing in mind changing human nature, from low faeeliya to high faeeliya, people will change and determine the nature of economy and the nature of state authority. In specific terms, what is needed is a change from control to participation in all levels. There's no doubt that participation in power, economy, and truth-making is the road map to postcapitalism society. Such a transition must be peaceful, democratic, and humane, reflecting the faeeliya of the new social actors.

CONCLUSION

Our objective is to give a coherent scientific explanation for the question of why living systems self-organize, self-reproduce, and evolve while nonliving systems don't. We find two conflicting opponents, reductionism and antireductionism, and neither of them has been able to resolve the problem based on its conception and argument. On the one hand, life is irreducible to ordinary physics because such reduction does not answer the question about why living systems self-organize and evolve while nonliving systems don't. On the other hand, life cannot be independent of physics because it shows no violation of physical laws, which means it depends on physics.

To overcome this dilemma, I propose broadening the ontological foundation of contemporary physical theory by discovering a new physical property that distinguishes life from nonlife, the property that has escaped human imagination for such a long time. However, to discover such a property, we need a paradigm shift; we need to extend our conception of information, DNA geometry, and quantum field and action principle. Based on the paradigm shift, we find that the secret of life resides on the DNA or genome as quantum information fractal field, which generates, in addition to weak electromagnetic waves, bioinformation oscillations for successive generations. The bioinformation oscillations contain the dynamical essence of living systems, so they can be represented by a generalized Schrödinger type of system, which we call the life-organizing principle. The life-organizing principle is a nonlinear, nonconservative, and irreversible system that admits limit cycle. In other words, it is an attractor. It is a minor attractor when we deal with cellular dynamics and major attractor when we deal with multicellular organism dynamics. A cell type is an example of a minor attractor that belongs to a basin of a major attractor.

Based on the life-organizing principle, the following principle and laws can be derived:

- Maximum action principle. The maximum action principle is the physical complement of the least action principle; according to the latter, if the passage of a conservative dynamic system from one configuration to another is spontaneous, the corresponding action has minimum value—that is, the rate of change of action decreases. On the contrary, according to the former, the rate of change of action increases.

- First law of self-organization: This law is based on the maximum action principle, and it states that the organism's rate of change of action increases in direct proportion to its developmental functional complexity (vitality or bioinformation).

- Second law of self-organization: The law states that the organism's total vitality (product of organism's total action, life span, and genome's physical information) is directly proportional to Fibonacci numbers and the genome physical information and inversely with the frequency of bioinformation oscillations. The Fibonacci numbers represent the quantum functionally stationary states; they represent minor attractors states ontogenetically and major attractors states phylogenetically. Based on the second law, both phylogeny and ontogeny are processes that generate and assemble minor attractors.

- We have also demonstrated that the genome's total bioinformation is conserved and generates two survival components: reproductive fitness component and total vitality fitness component.

Each of these laws is biologically universal, meaning applicable to all living systems phylogentically as well as ontogentically. Not only that, but also they admit limiting transition to linear reversible quantum mechanics when the organism is dead. So we claim that we have discovered a generalized physics or holistic science that facilitates a foundation for ordinary physics, biology, and human sciences. Holistic science is not a substitute for specialized sciences; equally well, the specialized sciences are not a substitute for holistic science. Holistic science is capable of resolving problems unsolvable based on specialized sciences, such as discovering links or bridges between the specialized sciences. It is also important and more productive that scientists who are increasingly immersed in extremely narrow specialty operate based on a holistic conception of scientific reality, thus broadening the concept of matter by discovering the bioinformation oscillations, bridging the gap between ordinary physics and biology.

Moreover, the discovery that the genome total bioinformation generates two survival components, reproductive fitness component and total vitality (faeeliya) fitness component, bridges the gap between biology and human sciences. Consequently, human nature being complementary, Darwinian, at low faeeliya, displays the Darwinian instincts of selfishness, violence, and despotism. However, a high faeeliya human nature, being faeelistic, displays the properties of altruism and creativity. The complementary nature of human nature resolves the existing contradiction between science and the final goals of religion. Because "altruistic groups beat selfish groups," the social need for faeeliya fitness creates morality and religion; however, religious discourse, being multilevel discourse, addresses all mind structures.

The sociocultural evolution as well as biological evolution being a maximization of faeeliya, humans can transcend the Darwinian phase and maintain postcapitalism society, not by historical determinism but by the determinism of their own faeeliya. The transition to postcapitalism requires a faeeliya system of knowledge, a faeeliya cultural project, faeeliya development, and a new social force that can actualize the project. The transition must be peaceful, democratic, and humane, reflecting the faeeliya of the new social actors.

How to Test the Theory?

The theory provides numerous predictions that can be accounted for experimentally to falsify the theory. Fortunately, the theory is quantitative and all the variables are measurable. If we consider biocomplexity, it is a product of the organism's maximum rate of change of action (which is equal to its metabolic rate at adulthood) times its genome's physical information. Therefore, we can compare the biocomplexity of different species and evaluate it with regard to other biocomplexity measures (e.g., anatomical, physiological, number of cell types, and so forth). Another important quantity is total vitality, which is a product of the organism's total action, life span, and genome's physical information. Total vitality is associated with numerous predictions. First, it is constant under constant environmental conditions, meaning that the organism's total bioinformation is constant so the organism is capable of reproducing its own kind. Second, total vitality is evolution target criterion; it measures evolutionary progress along a given lineage as an increase of total vitality. Third, the quantum stationary functional states are represented by Fibonacci's numbers. Fourth, ontogenetic development is also a process of increase of total vitality; in particular, cellular differentiation is a process of increase of total vitality—meaning increase of cell cycle time and/or increase of cell total action (body size). Fifth, both ontogeny and phylogeny are processes that generate and assemble minor attractors (cell types).

Sixth, for the genetic code to be meaningful (i.e., capable of generating viable functional proteins) it has to be structured along a path of maximum action, which means it has to be subject to the second law of self-organization. The structuring of nucleotides and codons must incorporate Fibonacci's numbers and the golden ratio.

APPENDIX

Proof of result 1

Since v(0) = E(0)A^a > 0, v(A) = 0 $\Rightarrow$ v(t) has a single maximum for $0 \leq t \leq A$, and we take this maximum at adulthood—when t = α so that $\dot{v}(\alpha) = 0$. This assumption determines the value of a:

$\therefore$ v (t) = bE(t) ℓ^a

$\therefore \dot{v}(t) = b\dot{E}\ell^a - abE(t)\ell^{a-1}$

$\therefore \dot{v}(\alpha) = 0 \Rightarrow a = \dfrac{\dot{E}(\alpha)\ell(\alpha)}{E(\alpha)}$ 　(105)

$\therefore \dot{v}(0) = b\dot{E}(0)A^a - abA^{a-1}E(0) > 0$

$\therefore \ddot{v}(\alpha) = b\ddot{E}(\alpha)\ell^a(\alpha) - 2ab\dot{E}(\alpha)\ell^{a-1}(\alpha) + ba(a-1)E(\alpha)\ell^{a-2}(\alpha) =$

$-b(a+1)\dot{E}(\alpha)\ell^{a-1}(\alpha) < 0$, 　Where $\ddot{E}(\alpha) = 0$

Consequently, in this model, there exists a vitality function v(t) that satisfies the following condition:

It increases before adulthood, has a maximum at adulthood, decreases afterward, and becomes zero when the organism dies.

Note: The life span A is an average value, so some organisms die for different reasons for t < A; others survive for t > A.

Numerical example for a vitality curve:

To substantiate result 1 further, we introduce dimensionless representation of the vitality curve using the exponential energy growth functions of some classes of organisms (Medawar 1945; Bertalanffy 1957). Thus we can deduce that the energy growth function, E(t), for some classes of organisms is given by:

$$E(t) = ge^{Rt} \quad t \le \alpha \quad (106)$$

Therein, g and R are positive constants that depend on the species of the organism. And α is the time or age of the organism, when it is fully grown, (i.e., adult). For the period $\alpha \le t \le A$, we make the simple extension that the metabolic rate $\dfrac{dE}{dt}$ remains approximately constant at its value when $t = \alpha$.

Therefore:

$$\begin{aligned}
E(t) &= ge^{R\alpha} + (t - \alpha)\, gRe^{R\alpha} \\
&= ge^{R\alpha} \{1 + R(t - \alpha)\} \quad \alpha \le t \le A
\end{aligned}$$

$$v(t) = ge^{Rt} (A-t)^a \qquad\qquad t \le \alpha$$

$$(107)$$

$$\dot{v}(t) = ge^{Rt} (A-t)^{a-1} \{R(A-t)-a\} \qquad t \le \alpha \quad (108)$$

$$v(t) = ge^{R\alpha} \{1 + R(t-\alpha)\}(A-t)^a \qquad \alpha \le t \le A$$

Then v(A) = 0 and v(t) has a single maximum for $0 \le t \le A$.

To restrict our model further, we shall assume that v(t) takes this maximum value at adulthood—that is, when $t = \alpha$ so that $\dot{v}(\alpha) = 0$. This assumption determines the value of a:

$$a = R(A - \alpha) \quad (109)$$

Note that a > 0 as stated earlier since R > 0 and $A - \alpha > 0$.

Consequently, using dimensionless variables, we can draw the vitality curve. Let

$$Rt = x, \qquad AR = c, \qquad R\alpha = c_0 \quad (110)$$

and

$$v(t) = \frac{gy}{R^a}, \qquad a = c - c_0, \qquad \text{from eqn (109)} \quad (111)$$

$$y(x) = e^x R^a (A - \frac{x}{R})^a = e^x (AR - x)^a = e^x (c - x)^a \quad (112)$$

$$x \leq R\alpha$$
$$x \leq c_0$$

From (108) :

$$\frac{gy(x)}{R^a} = ge^{c_0} \{1 + R(\frac{x}{R} - \alpha)\} (A - \frac{x}{R})^a$$

$$y(x) = R^a e^{c_0} (1 + x - c_0) \frac{1}{R^a} (c - x)^a$$

Hence,

$$= e^{c_0} (1 - c_0 + x)(c - x)^a$$

$$\alpha \leq \frac{x}{R} \leq A$$
$$c_0 \leq x \leq c$$

$$y(x) = \begin{cases} e^x(c-x)^{c-c_o} & x \le c_o \\ e^{c_o}(1-c_o+x)(c-x)^{c-c_o} & c_o \le x \le c \end{cases}$$

$$c > c_o, \quad \text{take} \quad c = 8 \quad \text{and} \quad c_o = 4 \tag{113}$$

Thus we get table 1 and figure 1. It is worth mentioning that figure 1 is based on a particular form for the energy growth function, equations (106) and (107). However, it is evident that no matter what energy growth function the organism has, the present model determines its vitality curve.

REFERENCES

Adami, C. 2002. "What Is Complexity?" *BioEssays* 24(12): 1085-1094.

Alberts, B. Johnson, A. Lewis, J. Ra, M. Roberts, K. and Walter. P. 2002. *Molecular Biology of the Cell.* 4th ed. New York and London. Garland Science.

Alexander, G. Douglas. 1970. New York and London: *Biology.* Barnes and Nobel Books.

Antoniou, I., and I. Prigogine. 1993. "Intrinsic Irreversibility and Integrability of Dynamics." *Physica A* 192: 443-464.

Asimov, I. 1972. *The Biological Sciences.* Vol. 2 of *Asimov's Guide to Science.* Harmondsworth, Middlesex, England: Penguin Books Ltd.

Ayala, J. F. 1977. "Philosophical Issues." In *Evolution*, edited by Dobzhansky, 511. San Francisco: W. H. Freeman and Company.

Ayala, J. F. 1983. "Micro-evolution and Macro-Evolution." In *Evolution from Molecules to Men*, edited by S. D. Bendall, 391, 396. Cambridge University Press.

Bak, P., and K. Sneppen. 1993. "Punctuated Equilibrium and Criticality in a Simple Model of Evolution." *Physical Review Letters* 71: 4083-4086.

Bak, P., and M. Paczuski. 1995. "Complexity, Contingency, and Criticality." *Proceedings of the National Academy of Sciences USA* 92: 6689-6696.

Bedau, Mark A. 2007. "What Is Life?" http://people.reed. edu/~mab/publications/papers/bedau_ch24_Blackwell.pdf.

Bertalanffy, L. Von. 1954. *Problems of Life.* New York: Academic Press.

Bertalanffy, L. Von. 1957. "Quantitative Laws in Metabolism and Growth." *Quarterly Review of Biology* 32: 217-231.

Bohr, N. 1933. "Light and Life." *Nature* 131: 457-459.

Bonner, J. T. 2004. "Perspective: The Size-Complexity Rule." *Evolution* 58 (9): 1883-1890.

Brooks, D. R. 2001. *Evolution in the Information Age: Rediscovering the Nature of the Organism. Semiosis, Evolution, Energy, Development.* Vol. 1

Brooks, D. R., and E. O. Wiley. 1986. *Evolution as Entropy: Toward a Unified Theory of Biology.* Chicago: University of Chicago Press.

Brooks, D. R., and E. O. Wiley. 1988. *Evolution as Entropy: Toward a Unified Theory of Biology.* Chicago: University of Chicago Press.

Bryant, P., and P. Simpson. 1984. "Intrinsic and Extrinsic Control of Growth in Developing Organs." *Quarterly Review of Biology* 59 (4): 387-415.

Chaisson, J. E. 2005. "Follow the Energy: Relevance of Cosmic Evolution for Human History." *Journal of Historical Society* 6 (5): 26.

Collier, J. 2003. "Hierarchical Dynamical Information Systems with a Focus on Biology." *Entropy* 5: 100-124.

Conlon, and M. Raf. 1999. "Size Control in Animal Development." *Cell* 96 (2): 235-244.

Davies, P. C. W. 2004. "Does Quantum Mechanics Play a Non-Trivial Role in Life?" *BioSystems* 78 (2004): 69-79.

Dawkins, R. 1986. *The Blind Watchmaker.* UK: Longman Scientific & Technical.

Dawkins, R. 1995. *At Home in the Universe.* Oxford: Oxford University Press.

Delbruck, M. 1949. "A Physicist Looks at Biology." *Transactions of the Connecticut Academy of Arts and Sciences* 38: 173-190.

Del Giudice, et al. 1993. "Hamiltonian and Superradiant Phase Transition." *Modern Physics Letters B* 7: 1851-1855.

Del Giudice, E., P. Stefanini, A. Tedeschi, and G. Vitiello. 2011. "The Interplay of Biomolecules and Water at the Origin of the Active Behavior of Living Organisms." *Journal of Physics: Conference Series* 329

Dembski, W. 1998. "Science and Design." *First Things* 86, (October *1998*): 21-27

Dobzhansky, Th. 1955. *Evolution, Genetics, and Man.* New Delhi: Wiley Eastern Private Limited.

Dobzhansky, Th. 1977. *Evolution.* San Francisco: W. H. Freeman and Company. 17.

Draganescu, M. "Deep Reality, Conscious Universe, and Complementarity." http://www.racai.ro/~dragam/2KAFN ADO.HTM.

Eden, M. 1966. "Inadequacies of New Darwinian Evolution as a Scientific Theory." In *Mathematical Challenges to the New-Darwinian Interpretation of Evolution,* edited by S. P. Moorhead and M. M. Kaplan. Philadelphia. The Wistar Institute Press.

Eign, M. 1992. *Steps toward Life—A Perspective on Evolution.* Oxford: Oxford University Press.

Eldredge, N., and J. S. Gould. 1972. "Punctuated Equilibria: An Alternative to Phyletic Gradualism." In *Models in Paleobiology,* edited by T. J. M. Schopf, 85-115. San Francisco: Freeman, Cooper.

Eldredge, N. 1985. *Unfinished Synthesis: Biological. Hierarchies and Modern Evolutionary Thought.* Oxford: Oxford University Press. 6.

Elssaser, W. 1981. "Principles of New Biological Theory." *Journal of Theoretical Biology* 89: 131-150.

Elsheikh, M. 1988. Evolution of Unicellular Organisms in Terms of Vitality Function. *Trilogia* (Chile), Numero especial, Junio. Santiago, Chile.

Elsheikh, M. 1999. "Evolution: A Nonlinear Irreversible Quantum Model." *Altahadi University Scientific Journal.* *Vol.1 (1).*

Elsheikh, M. 2005. "Toward a Science of Faeeliya." *Consequentiality*, volume 1 of *Human all too human. Expanding Human Consciousness*, edited by Dena Hurst. Tallahassee, Florida.

Elsheikh, M. 2010. "Toward a New Physical Theory of Biotic Evolution and Development." *Ecological Modelling* 221: 1108-1118.

Elsheikh, M. www.lifeprinciple.com. Last modified November, 2013.

Fath, B., S. Jorgensen, B. Patten, and M. Straskraba. 2004. "Ecosystem Growth and Development." *BioSystems* 77: 213-228.

Frigg, R. 2003. "Self-Organised Criticality—What It is and What It Isn't." *Studies in History and Philosophy of Science* 34: 613-632.

Galtin, L. L. 1972. *Information Theory and the Living System.* New York: Columbia University Press.

Ganti, T. 2003. *The Principles of Life* (with a commentary by James Grisemer and Eörs Szathmáry). New York: Oxford University Press.

Fox, W. S. 1984. *Proteinoid Experiments and Evolutionary Theory.* In M.W. Ho and P. T. Saunders, Eds., *Beyond Neo-Darwinism.* London: Academic Press, pp. 15-60.

Gitt, W. 1996. "Information, Science, and Biology." *Technical Journal* 10(2): 181-187.

Goodwin, C. B. 1963. *Temporal Organization in Cells: A Dynamic Theory of Cellular Control Process.* London and New York: Academic Press.

Goodwin, C. B. 1985. "What Are the Causes of Morphogenesis?" *BioEssays* 3: 32-36

Goodwin, C. B. 1984. "Field Theory of Reproduction and Evolution." In *Beyond Darwinism*, edited by Mae-Wan HO and T. P. Saunders, 230. London and New York: Academic Press.

Gould J. S. 1983. "The Changing Role of Paleontology." In *Evolution from Molecules to Men*, edited by S. D. Bendall, 362-363. New York: Cambridge University Press.

Gould, J. S. 1982. "Darwanism and the Expansion of Evolutionary Theory." *Science* 21: 380-387.

Grandpierre, A. 2007. "Biological extension of the action principle." *NeuroQuantology* 4: 346-367.

Grant, P. 1978. *Biology of Developing Systems.* New York and London: Holt-Sanders International Editions.

Ho, M. W. 1995. "Bioenergetics and the Coherence of Organisms." *Neuronetwork World* 5: 733-750.

HO, M. W., and T. P. Saunders, eds. *Beyond Neo-Darwinism.* London: Academic Press. 19.

Hut, P., B. Goodwin, and S. Kauffman. 1997. "Complexity and Functionality: A Search for the Where, the When, and the How." Proceedings of the International Conference on Complex Systems, Nashua, NH.

Jorgensen, S. E. 2006. "Toward a Thermodynamics of Biological Systems." *International Journal of Ecodynamics* 1 (1): 1-19.

Jorgensen, S. E. 2006-2007. *An Integrated Ecosystem Theory.* Ann.—Eur. Acad. Sci. Liège: EAS Publishing House, 19-33.

Jorgensen, S. E., and B. D. Fath. 2007. *A New Ecology Systems Perspective.* Elsevier: Linacre House, Jordan Hill, Oxford OX2 8DP, UK Radarweg 29, PO Box 211, 1000 AE Amsterdam, The Netherlands

Jorgensen, S. E. 2007. "Evolution and Exergy." *Ecological Modelling.* doi: 10. 1016/j.ecolmodel.2006.12.035

Kaila, V.R.I, A. Annila. 2008, "Natural Selection for Least Action." *Proceedings of the Royal Society. A.*

Mathematical, physical and engineering sciences, vol 464, no. 2099 : 3055-3070.

Kamshilov, M. M. 1976. *Evolution of the Biosphere.* Moscow: MIR Publishers, 200.

Kauffman, S. 1995. *At Home in the Universe.* Oxford: Oxford University Press.

Kim, J. 1998. *Mind in a Physical World: An Essay on the Mind-Body Problem and Mental Causation.* Representation and Mind Series. Cambridge: A Bradford Book, The MIT Press.

Kleiber, M. 1947. "Body Size and Metabolic Rate." *Physiological Reviews* 27: 511-541.

Knott, R. 2001. "The Fibonacci Numbers and Golden Section in Nature." http:/www.mcs.Surrey.ac.uk./personal/R.Knott/Fibonacci/Fibnat.html.

Kurakin, A. 2011. "The Self-Organizing Fractal Theory as a Universal Discovery Method: The Phenomenon of Life." *Theoretical Biology and Medical Modelling.*

Krylov, N., and N. Bogoliubov. 1947. *Introduction to Nonlinear Mechanics.* Princeton, New Jersey: Princeton University Press.

Laszlo, E. 2004. "Nonlocal Coherence in the Living World." *Ecological Complexity* 1: 7-15.

Lifson, S. 1997. "On the Crucial Stages in the Origin of Animate Matter." *Journal of Molecular Evolution* 44 (1): 1-8.

Loboda O., and V. Goncharuk. 2010. "Theoretical Study on Icosahedral Water Clusters." *Chemical Physics Letters* 484: 144-147.

Longo, G., M. Montevil, and S. Kauffman. 2012. "No Entailing Laws, but Enablement in the Evolution of the Biosphere." *arXiv* : 1201.2069v1 [q-bio. OT] 10.

Loutrup, S. 1984. "Ontogeny and Phylogeny." In *Beyond Neo-Darwinism*, edited by Mae-Wan Ho and T. P. Saunders, 169. London: Academic Press.

Lloyd, S. 2002. "Computational Capacity of the Universe." *Phys Rev Letters* 88, 237901-237908.

Mandelbrot, B. 1982.*The Fractal Geometry of Nature*. San Francisco: W. H. Freeman and Company.

Maynard Smith, J., and E. Szathmary. 1999. *The Origins of Life*. Oxford: Oxford University Press.

McClendon J. H. 1980. "The Evolution of the Chemical Isotopes as an Analog of Biological Evolution." *J. Theor. Biol.* 87: 113-28.

Medawar, P. B. 1945. *Essays on Growth and Form*, edited by C. Gross and P. B. Medawar. Oxford: Oxford University Press.

Meyer, S. C. 2004. "The Origin of Biological Information and the Higher Taxonomic Categories." *Proceedings of the Biological Society of Washington* 117 (2): 213-239.

Migulin, V. 1983. "Self-Organization in Systems with One Degree of Freedom." In *Basic Theory of Oscillations*, edited by Mugulin, 184. Moscow: Mir Publishers.

Montagnier L., J. Aissa, S Ferris, and C. Lavallee. 2009. DNA waves and Water. *Interdiscip. Sci. Comput. Life Sci.* 1: 81-90

Montagnier L., J Aissa, E. Del Giudice, C Lavallee, A Tdeschi, and G. Vitiello. 2010. "DNA Waves and Water." arxiv.org/abs/1012.5166.

Nicolis, G., and I. Prigogine. 1977. *Self-Organization in Nonequilibrium Systems: From Dissipative Structures to Order Through Fluctuations.* New York and London: Wiley-Interscience Publication, John Wiley & Sons.

Niklas, J. K. 2006. "Plant Allometry, Leaf Nitrogen and Phosphorous Stoichiometry, and Interspecific Trends in Annual Growth Rates." *Ann Bot.* 97 (2): 155-163.

Odum, E. P. 1983. *System Ecology.* New York: Wiley Interscience.

Okasha, S. 2005. Multilevel Selection and the Major Transitions in Evolution. *Biol. and Philos.* 20: 989-1010.

Pareto, V. 1897. *Cours D'economie Politique*, republished as *Manual of Political Economy*. New York: Augustus M. Kelley Publishers.

Pattee, H. 1968. "The Physical Basis of Coding and Reliability." In *Toward a Theoretical Biology*, edited by Waddington H. Edingburgh University Press.

Pattee, H. 1971. "Can Life Explain Quantum Mechanics?" In *Quantum Theory and Beyond*, edited by Ted Bastin. Cambridge University Press.

Perez, Jean-Claude. 2010. "Codon Populations in Single-Stranded Whole Human Genome DNA Are Fractal and Fine-Tuned by the Golden Ratio 1.16." *Interdesci Sci Corput Life* 2: 228-240.

Preparata, G. 1995. *QED Coherence in Matter.* New Jersey, London, Singapore, and Hong Kong: World Scientific.

Prinzinger, R. 2005. "Programmed Aging: The Theory of Maximal Metabolic Scope." *Science & Society*. EMBO reports 6/special issue.

Pross A. 2003. "The Driving Force for Life Emergence: Kinetic and Thermodynamics Considerations." *J. Theor. Bio.* 220: 293-406.

Ray, T. 1992. An Approach to the Synthesis of Life. In *Artificial Life II*, edited by C. Langton, C. Taylor, J. D. Farmer, and S. Rasmussen, 371-408. Redwood City, California: Addison-Wesley.

Root-Bernstein, R. S., and P. F. Dillon. 1997. "Molecular Complementarity 1: The Complementarity Theory of the Origin and Evolution of Life." *J. Theor. Biol.* 188: 447-479.

Rosen, R. 1970. *Dynamical System Theory in Biology.* New York, London: Wiley Interscience.

Sabelli, H. 2008. "Bios Theory of Innovation." *The Public Sector Innovation Journal* 13 (3): art. 12.

Saunders, T. P. 1984. "Development and Evolution." In *Beyond Neo-Darwinism*, edited by Mae-Wan HO and T. P. Saunders, 243-245. London: Academic Press.

Scaruffi, P. 2003. "Thinking about Thought". iUniverse Publisher, Bloomington, USA

Schrödinger, E. 1944. *What Is Life?* Cambridge: Cambridge University Press. 85-86.

Shackney, E. S. 1973. "A Cytokinetic Model for Heterogeneous Mammelian Cell Population." *J. Theor. Bio.* 38: 305-333.

Schweickart, D. 2011. *After Capitalism.* New York: Rowman & Littlefield.

Sewell, G. 2011. "A Second Look at the Second Law." *Applied Mathematics Letters.* Elsevier.

Smith, J. D. H. 1999. "On the Evolution of Semiotic Capacity." In *Semiotics, Evolution, Energy*, editd by E. Taborsky, 283-309. Shaker Verlag, Aachen.

Sneppen, et al. 1995. "Evolution as a Self-Organized Critical Phenomenon." *Proc. Natl. Acad. Sci. USA* 92: 5209-5213.

Stansfield, D. W. 1977. *The Science of Evolution.* New York: MacMillan Publishing Co. 27.

Thomas, G., H. Sabelli, L. Kauffman, and L. Kovacevic. 2006. "Biotic Patterns in the Schrödinger Equation in the Early Universe." *International Journal of Complex Systems in Science.*

Thomson, K. S. 1992. "Macroevolution: The Morphological Problem." *American Zoologist* 32: 106-112.

Ulanowicz, R. E. 2004. "Quantitative Methods for Ecological Network Analysis." *Computational Biology and Chemistry* 28: 321-339.

Van Regenmortel, M. H. V. 2004. "Reductionism and Complexity in Molecular Biology." *EMBO reports* 5 (11).

Varute, T. A., and S. K. Bahatia. 1976. *Cell Structure and Function.* New Delhi: Vicas Publishing House Pvt. Ltd.

Vicente, A. 2006. "On the Causal Completeness of Physics." *International Studies in the Philosophy of Science* 20: 149-171.

Voit, O. E., and G. Dick. 1983. "Growth of Cell Populations with Arbitrary Distributed Cycle Durations." *Mathematical Biosciences* 66: 247-262.

Waddington, C. H. 1968. "Does Evolution Depend on Random Search?" In *Toward a Theoretical Biology*, edited by Waddington. The Kynock Press Birmingham.

Waddington, C. H. 1968. "The Basic Ideas of Biology." In *Toward a Theoretical Biology 1: Prolegomena*, edited by Waddington. Edinburgh University Press.

White, M. 2008. "What Is a Dodecahedron?" Rafiki, Inc.

Whyte, L. L. 1965. "Internal Factors in Evolution". Associated Book Publishers. London E.C.4

Wille, J. J. 2012. "Occurrence of Fibonacci Numbers in Development and Structure of Animal Forms: Phylogenetic Observations and Epigenetic Significance." *Natural Science* 4: 216-232.

Wilson, D. S., and E. O. Wilson. 2007. "Rethinking the Theoretical Foundation of Sociobiology." *The Quarterly Review of Biology* 32 (4).

Winter, D. http://www.fractalfield.com/mathematicsoffusion/.

Winter, D. http://www.goldenmean.info/fractalfield/.

Yates, E. F. 1982. "Outline of a Physical Theory of Physiological System." *Can. J. Physio. Pharmacal.* 60.

Yockey, H. P. 1992. *Information Theory and Molecular Biology*, Cambridge: Cambridge University Press. 255-257.

Zotin, A. A. 2006. "Equations Describing Changes in Weight and Mass-Specific Rate of Oxygen Consumption in Animals during Postembryonic Development." *Biology Bulletin* 33: 323-331.

.